T0133180

MULTICULTURAL SCIENCE
IN THE
OTTOMAN EMPIRE

DE DIVERSIS ARTIBUS

COLLECTION DE TRAVAUX
DE L'ACADÉMIE INTERNATIONALE
D'HISTOIRE DES SCIENCES

COLLECTION OF STUDIES
FROM THE INTERNATIONAL ACADEMY
OF THE HISTORY OF SCIENCE

DIRECTION
EDITORS

EMMANUEL
POULLE

ROBERT
HALLEUX

TOME 69 (N.S. 32)

BREPOLS

MULTICULTURAL SCIENCE
IN THE
OTTOMAN EMPIRE

Ekmeleddin IHSANOGLU, Kostas CHATZIS,
Efthymios NICOLAIDIS

BREPOLS

The XXth International Congress of History of Science was organized by the Belgian National Committee for Logic, History and Philosophy of Science with the support of :

© 2003 Brepols Publishers n.v., Turnhout, Belgium

D/2003/0095/66
ISBN 2-503-51446-4
Printed in the E.U. on acid-free paper

TABLE OF CONTENTS

Introduction
Science, technology and cultural diversity : from the Ottoman Empire
to the National States..7
    Ekmeleddin IHSANOGLU and Efthymios NICOLAIDIS

Early Modern Ottoman and European gunpowder technology 13
    Gábor ÁGOSTON
Un Alchimiste vagabond en Europe et au Proche Orient
durant le XVIIe siècle...29
    Ioli VINGOPOULOU
The Confrontation of mathematics on behalf of the Eastern Orthodox
Church during the Ottoman period 53
    Maria TERDIMOU
" Renegades " and missionaries as minorities in the transfer
of knowledge..63
    Sonja BRENTJES
Ottoman engineer Mehmed Said Efendi and his works on a geodesical
instrument (*Müsellesiye*)...71
    Mustafa KAÇAR and Atilla BIR
Ottoman engineer Mehmed Said Efendi and his treatise
on vertical sundial..91
    Atilla BIR and Mustafa KAÇAR
Scientific practice, patronage, salons, and enterprise in eighteenth century
Cairo : examination of al-Gabartī's history of Egypt 107
    Rainer BRÖMER
Éducation et politique au XIXe siècle : Les élèves Grecs dans les grandes
écoles d'ingénieurs en France 121
    Fotini ASSIMACOPOULOU et Konstantinos CHATZIS
Crossing communal boundaries : technology and cultural diversity
in the 19th century Ottoman Empire...139
    Yakup BEKTAŞ

Unity and diversity on Ottoman railways : a preliminary report on technology transfer and railway workers in the Ottoman Empire 149
    Peter MENTZEL

Orient et Occident, lectures d'une polarité scientifique et technique 165
    Antoine PICON

Faire l'histoire des usages des objets techniques : formules, projets et pratiques. L'exemple des brouettes sur le chantier du Canal de Suez (1859-1869) ... 173
    Nathalie MONTEL

Attitudes, activities and achievements : science in the modern Middle East 181
    Yakov M. RABKIN

Beyond culturalism ? An overview of the historiography on Ottoman science in Turkey 201
    Cemil AYDIN

The state and professional identities : the emergence of the socio-professional class of Greek engineers at the beginning of the 20th century 217
    Yiannis ANTONIOU and Michalis ASSIMAKOPOULOS

Institutionalisation of science education and scientific research in Turkey in the 20th century 235
    Sevtap KADIOGLU

Contributors 247

INTRODUCTION

SCIENCE, TECHNOLOGY AND CULTURAL DIVERSITY : FROM THE OTTOMAN EMPIRE TO THE NATIONAL STATES

Ekmeleddin IHSANOGLU and Efthymios NICOLAÏDIS

During the last two decades the international history of science congresses organized by the International Union of History and Philosophy of Science (IUHPS) have devoted increasing attention to the history of science and technology in non-western cultures, particularly in the Islamic world. This attention was institutionalized at the XVIIIth International Congress in Hamburg in 1989 with the establishment of an International Commission on Science and Technology in Islamic Civilization within the IUHPS.

Science in Islamic civilization, which emerged as the common product of various scholars who inherited the scientific legacy of different pre-Islamic civilizations flourished from the 8th until the 16th centuries, became the subject matter of several general studies and research works. However, the sub-cultures that constitute this civilization and particularly science in the later period of Islamic civilization, which may be considered as a distinct chapter of these scientific activities, have not been sufficiently studied yet.

However, the scientific activities of the Ottoman Empire have begun to receive the attention of a greater number of historians of science who became interested in the following two-sided aspects of Ottoman science : Indeed Ottoman science was a continuation of the classical tradition of Islamic science, and Ottomans constituted the first non-western culture to come into contact with western science and to attempt to transfer it. As a great empire that ruled over vast lands in Asia, Europe and Africa between the years 1299-1923 the Ottoman Empire represented a cultural multiplicity resulting from the centuries of peaceful coexistence *(Pax Ottomanica)* of various ethnic and religious entities within it. This cultural diversity was reflected in the scholarly activities that took place within the Ottoman world. The first serious studies on history of Ottoman science started on an institutional basis in Istanbul around 1980s.

Institutions such as IRCICA, founded in 1980, University of Istanbul, Faculty of Letters, History of Science Chair (1984-2001), and TBTK (Turkish Society for History of Science), founded in 1989, organized national and international symposia jointly most of the time. These meetings had a significant role in presenting the findings of the studies on Ottoman science to the attention of historians of science all over the world. Following the publication of the proceedings of these symposia historians of science from various countries began to conduct studies in this field.

History of Ottoman Scientific Literature series, which started to be published in 1990 and consists of seven volumes on astronomy, mathematics, geography and music, presents information about the scholarly works produced in the Ottoman world and their authors. It is hoped that this series, prepared for publication by IRCICA, will continue to be published in separate volumes on military science, medical and natural sciences.

In parallel, during the last two decades appeared studies and research projects on the spreading of the " scientific revolution ", or " new science " to the European periphery. As far as it concerns the Ottoman Empire, the role of Greek-language education and books on the spreading of the " new science " to the Christian communities of this Empire has been studied, as well as the role of the Byzantine tradition on the scientific education of these communities. The conflicts between this tradition and the " new science " have been pointed out, the receptivity and the resistance to these new ideas and the role played by " new science " to the constitution of the new national states created by national communities of the Empire, has drawn the attention of researchers. On those themes, a series of Conferences have been organised and a series of books have been published by the Programme of History of Science of the National Hellenic Research Foundation. Attention has also been drawn on the general theme of the constitution of a common scientific culture in Europe, based on the spreading of the scientific ideas born during the 16th-18th centuries in western and central Europe. The book *L'Europe des sciences : constitution d'un espace scientifique*[1] has summarized the research on that field.

The first Symposium on " Science, Technology and Industry in the Ottoman World ", held in 1997 within the framework of the XXth International Congress of History of Science, attracted the attention of scholars in this field and added a new dimension to the above-mentioned studies. The proceedings of this congress have been published[2]. In the framework of the same Congress a Symposium has been organised on the transfer of the scientific knowledge from

1. M. Blay, E. Nicolaïdis (eds), *L'Europe des sciences : constitution d'un espace scientifique*, Paris, Seuil, 2001.

2. E. Ihsanoglu, A. Djebbar, F. Günergun (eds), *Science, Technology and Industry in the Ottoman World* : proceedings of the XXth International Congress of History of Science (Liège, 20-26 July 1997), volume VI, Turnhout (Belgium), Brepols, 2000 (De Diversis Artibus. Collection of Studies from the International Academy of the History of Science, t. 46, n.s. 9).

Europe to the other countries[3]. In the meantime, a common project between France, Greece and Turkey (LATTS/ENPC, INR/NHRF and IRCICA) has been established on the transfer of technology and science from Western Europe to the East Mediterranean during the 19th century[4]. As a first result of that cooperation and following the two above-mentioned Symposia of the XXth International Congress, a Symposium has been organized by E. Ihsanoglu, K. Chatzis and E. Nicolaïdis within the XXIth International Congress of History of Science which was held in Mexico City on July 8-14, 2001. The theme of the Symposium, which took place on 9 and 10 July, was : " Cultural Diversity : from the Ottoman Empire to the National States ".

Scientific activities in the Ottoman world comprise various scientific traditions, including the Islamic tradition inherited by the Ottoman Turks and carried on by Arabs, who were part of the Ottoman Empire, and then joined by European peoples, such as the Bosnians and the Albanians newly converted to Islam, as well as the tradition of different Christian peoples living in Anatolia and the Balkans, (e.g., the Greek Colleges where " new " science was taught), and the contributions of native Jewish scholars as well as those who emigrated from Andalusia. The Ottoman world had the necessary grounds for the interaction of all these different traditions. The Ottoman Empire held vast lands in Europe and, as a result of the Ottomans' contact with European science from the very early ages, this new scientific tradition spread in the Ottoman lands for the first time outside its own cultural environment where it originated.

The Ottoman Empire gave rise to 29 national states in Europe, the Middle East and North Africa. The most significant aspect of the scholarly activities in the Ottoman Empire is that the Ottomans both depended on the previously established Turkish-Islamic scientific tradition and at the same time they engaged in attempts to transfer the new technologies and sciences that developed in the Western world. Generally speaking, during the classical period Ottoman science developed within the cultural and intellectual circles that flourished in institutions related to science, education, health, justice, religion and the military. These institutions, the prototypes of which were found in pre-Ottoman Islamic states such as the Timurids, the Ilkhanids and the Seljukids, developed and gained a more definite structure during the Ottoman period. The first Ottoman observatory, founded by Takiyeddin al-Rasıd in Istanbul under the patronage of the reigning sultan had a particular importance. Observations were

3. C.A. Lértora Mendoza, E. Nicolaïdis, J. Vandersmissen (eds), *The Spread of the Scientific Revolution in the European Periphery, Latin America and East Asia* : proceedings of the XXth International Congress of History of Science (Liège, 20-26 July 1997), volume V, Turnhout (Belgium), Brepols, 2000 (De Diversis Artibus. Collection of Studies from the International Academy of the History of Science, t. 45, n.s. 8).

4. Greek and French partners have organised conferences on the theme of the transfer of Technology ; the first volume has been published, E. Nicolaïdis, K. Chatzis (eds), *Science, Technology and the 19th century State*, Athens, National Hellenic Research Foundation, 2000, 149 p., and the second is in print.

made in this institution between 1573-1580. Besides these institutions, which existed until the end of the empire, new western type educational institutions were founded particularly from the 18th century onwards. Thus, the Ottoman institutions acquired a western character. At first this development was limited to military and technical educational institutions. From the 19th century onwards, however, western influences were felt in civil education. Thus, before long the attempt to establish the first Ottoman university took place in 1846. During the same period new institutions of higher education were founded in order to teach modern medicine, military science, engineering and law. Thus, institutions, which fulfilled similar functions within the eastern and the western traditions, coexisted side by side. A vivid cultural milieu developed in the empire thanks to this diversity and particularly to the religious and secular schools founded by non-Muslims within their communities as well as the secondary schools established by the state and foreign missionaries.

The Orthodox Christian subjects of the Sultan were under the authority of the Patriarch as far as their education was concerned. Indeed, the Sultan did not interfere in educational or scientific matters of the Christian community. As a consequence, in the Empire were installed parallel educational systems, those of the Muslim subjects and those of Orthodox Christian subjects. For the latter, education was provided in Greek language (the language of Byzantium) by individual teachers or, mainly after the beginning of the 17th century, in the Colleges organized and sponsored by the Orthodox Christian communities of the Empire. These Colleges of secondary level education provided the way through which science was taught to these communities. Indeed, due to the traditional relations with Italy (an important Greek community was established in Venice after the conquest of Constantinople by the Ottomans, many Byzantine scholars fled to Italy after this conquest), the Christian subjects who wished and had the possibility to continue their studies at a university level, went, during the 17th and 18th centuries, to the University of Padova (a Venetian town). In order to facilitate the studies of Greek students at this University, Colleges for Greeks have been founded by Greek donators in Venice and Padova. Another College for Greeks had been founded by Pope Leon X in Rome in 1514, and addressed mainly to Greek Catholics.

Except a few scattered texts, the new science was then introduced to the Christian communities of the Empire, following the reform of the University of Padova in 1738, when Giovanni Poleni founded the " Teatro di filosofia sperimentale ". The Greek pupils of Poleni edited in Greek language manuals of physics based on well-known books of that period, such as those of Peter van Musschenbroek or Abbé Nollet.

The introduction of the new science went together with the ideology of modernization of the society put forward by the followers of the Enlightenment and later by those of Nationalism. The Greek revolution against the Ottoman Empire had as an ideal the constitution of a European National State. That means, among other things, implementation of educational reforms in order to strenghten the teaching of science on all levels and to introduce modern technology. One of the first acts of the independent Greek State was to found State secondary education, a University, a Military School and shortly afterwards a School of Arts which would later become a Technical University. But, although this had not anymore to do with the structures of the Ottoman Empite, a complex network of relations persisted until at least the beginning of the 20th century, as a great number of Greeks lived in regions which belonged to this Empire.

All these facts about the nature of Ottoman science and the complex network of scientific and educational relations of the various populations in the Ottoman Empire, as well as the relations between the science of this Empire and Europe, have drawn the attention of historians of science in recent years, especially those of the countries that originated from the Empire[5].

Technology transfer from Europe to the Ottoman world has a long history that has not been elaborately studied yet. This is also valid for the scientific relations between the different religions of the Ottoman Empire. The Symposium on " Science and Technology and Cultural Diversity : from Ottoman Empire to the National States " which was held within the framework of the international congress organized by IUHPS in Mexico City on 8-14 July 2001, dealt with these issues, and the contributions of scholars from various disciplines and backgrounds enriched the field. The present volume is based on the papers of that Symposium.

---

5. For an overview of the scientific-educational institutions and literature during the Ottoman period see the following articles by E. Ihsanoglu, 1. " Ottoman Educational and Scholarly-Scientific Institutions ", 357-516 ; 2. " The Ottoman Scientific-Scholarly Literature ", 517-603 in E. Ihsanoglu (ed.), *History of the Ottoman State, Society & Civilisation,* vol. II, Istanbul, IRCICA, 2002.

EARLY MODERN OTTOMAN
AND EUROPEAN GUNPOWDER TECHNOLOGY[1]

Gábor ÁGOSTON

Scholars of European military history have traditionally held the opinion that the Islamic empires were reluctant to use firearms and European military technology because these forms of weaponry had not been attested to in the time of the Prophet and therefore were regarded as innovations (*bid'a*) that were incompatible with Islam. Scholarship on *bid'a* and on Ottoman military technology has long challenged such claims[2]. The adoption of the infidels' military techniques and weapons could be considered as a good innovation, for their use considerably helped Muslim monarchs in their endevor to expand the realm of their respectives empires at the expense of the infidels. The pragmatism of the fourteenth- and fifteenth-century Ottoman rulers, the flexible understanding and practice of Islam among various social groups of the expanding early Ottoman state, as well as the fact that the Ottomans followed the most tolerant school of Muslim jurisprudence (the *hanafi madhhab*) made it easier to find the adequate means of violence, and — if required — evaluate the necessary ideology to legitimize them. In the mid-sixteenth century Ogier Ghiselin de Busbecq, imperial ambassador to sultan Süleyman (r. 1520-66) drew his contemporaries' attention to the readiness of the Ottomans to adopt some useful inventions, and their reluctance to do the same in cases that they considered dangerous or harmful to their culture and society. In his frequently quoted passage he stated that " no nation in the world has shown greater readiness than

---

1. An abridged and somewhat different version of this paper will appear in a volume in honour of Professor Halil Inalcık.

2. Ignác Goldziher, *Muslim Studies*, II, London, 1967, 3 ; more recently see, Vardit Rispler, " Towards a New Understanding of the Term bid'a ", *Der Islam. Zeitschrift für Geschichte und Kultur des islamischen Oriens*, 68, 2 (1991), 320-328 ; on Ottoman attitudes towards Western technology see, Rhoads Murphey, " The Ottoman Attitude towards the Adoption of Western Technology : The Role of the Efrencî Technicians in Civil and Military Applications " in J.-L. Bacqué-Grammont, P. Dumont (eds), *Contributions à l'histoire économique et sociale de l'Empire ottoman*, Leuven, Eds. Peeters, 1983, 289-298, and *idem, Ottoman warfare, 1500-1700*, New Brunswick, New Jersey, Rutgers University Press, 1999, 13-15.

the Turks to avail themselves of the useful inventions of foreigners, as is proved by their employment of cannon and mortars, and many other things invented by Christians. They cannot, however, be induced as yet to use printing, or to establish public clocks, because they think that the scriptures —that is, their sacred books — would no longer be scriptures if they were printed, and that, if public clocks were introduced, the authority of their muezzins and their ancient rites would be thereby impaired "[3].

It is obvious from the evidence at our disposal that the Ottomans were far from being prisoners of the " extreme conservatism of Islam " — as suggested by the traditional Eurocentric secondary literature[4]. On the contrary, they were quick to adopt western military technology and know-how, and did so with a remarkable thoroughness. The adoption of European military technology was not a problem for the Ottomans of the fifteenth and sixteenth centuries. The sultans hired European military experts, imitated and perfected European weaponry, and at the end of the sixteenth century, when the Ottomans realized that the Europeans outperformed the sultan's soldiers in the use of hand-held firearms in the Hungarian front the Grand Vizier and a clear-sighted observer both advocated for the use of firearms, encouraged the introduction of adequate countermeasures in tactics and the restructuring of the Ottoman army[5]. The widespread generalization about the Muslim religious establishment's perceived hostility towards western technology and institutions likewise ought to be treated with caution. A group of Ottoman religious scholars who supported the military reforms during the eighteenth and nineteenth centuries, for instance, elaborated a special ideology, the so-called theory of reciprocity (*mukabele bi'l-misl*), according to which the use and adoption of the enemy's modern weapons was allowed if such weapons were to be used to defeat the enemy[6]. It is clear from historical evidence that the sultans had no difficulty to find religious scholars who supported their decisions with regard to the adoption of new weapons and military techniques.

3. Charles Thorrton Forster, F.H. Blackburne Daniell, *The Life and Letters of Ogier Ghiselin de Busbecq Seigneur of Bousbecque Knight, Imperial Ambassador*, I, London, 1881, 255.

4. Kenneth Meyer Setton, *Venice, Austria, and the Turks in the seventeenth century*, Philadelphia, PA, American Historical Society, 1991, 6, 100 and 450. For its critique see, Rhoads Murphey's review in *Archivum Ottomanicum*, XIII (1993-1994), 371-383.

5. Vernon J. Parry, " La Maniere de combattre ", in V.J. Parry and M.E. Yapp (eds), *War, Technology and Society in the Middle East*, London, Oxford University Press, 1975, 224 ; Halil Inalcık, " The Socio-Political Effects of the Diffusion of Fire-arms in the Middle East ", *War, Technology and Society*, 199 ; Mehmet İpşirli (ed.), " Hasan Kafı el-Akhisari ve Devlet Düzenine ait Eseri Usulü'l-hikem fi Nizamü'l-alem ", *Istanbul Üniversitesi Tarih Enstitüsü Dergisi*, 10-11 (1979-1980), 268 ; Karácson Imre, *Az egri török emlékirat a kormányzás módjáról*, Budapest, 1909, 20 ; G. Ágoston, " Az európai hadügyi forradalom és az oszmánok ", *Történelmi Szemle*, XXXVI, 4 (1995), 465-485. and *idem*, " Habsburgs and Ottomans : defense, military change and shifts in power ", *Turkish Studies Association Bulletin*, 22. 1., Spring ,1998, 126-141.

6. Uriel Heyd, " The Ottoman 'Ulema' and Westernization in the time of Selim III and Mahmud II ", *Islamic History and Civilization*, Uriel Heyd (ed.), Jerusalem, 1961, 74 (Scripta Hierosolymitana. Publications of the Hebrew University, IX).

Furthermore, when explaining opposition towards new military techniques on the part of the Ottoman religious establishment or certain military units, one has to take into consideration that opposition towards new weaponry was a general phenomenon in almost all societies. The introduction of new techniques and weaponry required the reorganization and restructuring of the armed forces that often had serious social and economic consequences. These real or perceived consequences in turn could often lead to resistance and objection on the part of those social groups who either did not fit any more into the reorganized structure or whose status within the army, and in the society at large, was negatively effected by the changes. In other words, instead of labeling any opposition towards non-Muslim technology (or towards military reforms in general) in the Ottoman context as an act that originated from "Islamic conservatism" (a term itself difficult to define), one should look at the social risks and consequences, and should do it in a comparative framework.

THE QUESTION OF INTRODUCTION OF GUNPOWDER WEAPONS IN THE OTTOMAN EMPIRE

Historians are generally obsessed with the "first" references to new technological devices, and the story of gunpowder and firearms is certainly not an exception. Although it is well-known that after the first appearance of gunpowder weapons, it took decades for armies to employ them regularly and in large enough quantities to be tactically significant[7], national histories usually pay special attention to the "first" dates when these weapons were supposedly introduced to their respective countries. Like their European colleagues, Turkish historians, too, tend to ascribe disproportionate importance to the "first", though dubious, references to firearms. The date of the introduction of firearms into the Ottoman Empire, however, remains debatable. The main difficulty, as with early Ottoman history in general, is the scarcity and questionable reliability of the sources. We hardly have any Ottoman sources contemporaneous with or close to the supposed time when firearms were introduced into the empire. It would be risky to base the argument solely on Ottoman chronicles dating from the late fifteenth and even the sixteenth centuries, that are not confirmed by other independent sources, since these chroniclers might have projected the terminology of their own times when referring to earlier events. Thus, in writ-

---

7. On this see, Carlo M. Cipolla, *Guns, Sails and Empires. Technological Innovation and the Early Phases of European Expansion 1400-1700*, London, 1965 (Reprint New York, Barnes and Noble, 1996); John F. Guilmartin, *Gunpowder and Galleys, Changing Technology and Mediterranean Warfare at Sea in the Sixteenth Century*, Cambridge, Cambridge University Press, 1974; William H. McNeill, *The Pursuit of Power. Technology, Armed Force, and Society since A.D. 1000*, Chicago, The University of Chicago Press, 1982; Geoffrey Parker, *The Military Revolution. Military Innovation and the Rise of the West, 1500-1800*, Cambridge, Cambridge University Press, 1988 (3rd ed. 1999); Kelly DeVries, *Medieval Military Technology*, Peterborough, Ontario, Broadwiew, 1992, and Bert S. Hall, *Weapons and Warfare in Renaissance Europe*, Baltimore, London, Johns Hopkins University Press, 1997.

ing about fourteenth-century sieges and battles involving only mechanical artillery (e.g., trebuchet) or personal missile weapons (e.g., crossbows) Ottoman chroniclers might erroneously mention firearms, which were used regularly at the time they wrote their annals.

Terminology constitutes a major problem for the student of Islamic military technology. As with many terms with regard to European and Byzantine military technology[8], in the Ottoman context, too, old terms were applied to new weapons or terms might have had multiple meanings. In fifteenth-century Ottoman sources the Turkish term " *top* " was used for both the shots of cannon and the cannon itself, and it is not always obvious which of these meanings we should apply. On the other hand, the mere lack of the term " cannon " or " gun " in the sources does not necessarily indicate that the weapon itself did not exist. (While the name saltpeter, for example, appears in English sources as late as the sixteenth century, saltpeter had been known and used two hundred years earlier).

We should bear this in mind since Turkish historians tend to ascribe too much importance to the first, though dubious, references to firearms cited in the earliest Ottoman chronicles. According to the standard historical chronology of the Ottoman Empire, the Ottomans cast a cannon as early as 1364 and used it, together with hand firearms, against the Karamanids in 1386[9]. Although these early references to Ottoman firearms have found their way into Western scholarly literature through Carlo M. Cipolla's influential work on European artillery[10], we should not forget that these dates derived from a sixteenth-century chronicle written by Şikari, who died in 1584, that is, two hundred years after the events in question. Furthermore, one should not forget that whereas the term cannon (*top*) is to be found in the manuscript of Şikari's chronicle that was the property of Ismail Hami Danişmend, the author of the aforementioned historical chronology, it is missing from all the other known copies of Şikari's work, making a later *interpolatio* even more plausible.

Referring to Neşri (d. before 1520), the late-fifteenth-century Ottoman chronicler, Ismail Hakkı Uzunçarşılı claimed that the Ottomans used cannon at the battle of Kosovo in 1389[11]. A recent study argues that the Ottomans used

8. The Spanish word *espingarda* originally was used to designate a type of crossbow, then came to mean a type of hand cannon. Bert S. Hall, *Weapons and Warfare*, 129, for the English term " gun " *cf.*, 44. For similar problems with regard to terminology in Byzantium see Mark C. Bartusis, *The Late Byzantine Army : Arms and Society, 1204-1453*, Philadelphia, University of Pennsylvania Press, 1992, 336.

9. Ismail Hami Danişmend, *Izahlı Osmanlı Tarihi Kronolojisi*, I, Istanbul, n.d., 73.

10. C.M. Cipolla, *Gun, Sails and Empires, op. cit.*, 90.

11. See, *inter alia*, Ismail Hakkı Uzunçarşılı, *Osmanlı Devleti Teşkilatından Kapukulu Ocakları. II. Cebeci, Topçu, Top Arabacıları, Humbaracı, Lağımcı Ocakları ve Kapukulu Suvarileri*, Ankara, Türk Tarih Kurumu, 1944, 35 (second edition 1984) ; Halil Inalcık, " Osmanlılar'da Ateşli Silahlar, " *Belleten*, 83. XXI, 509.

cannon as early as 1354 during the siege of Gallipoli. Yet again, this was based on a much later source, the early sixteenth-century chronicle of Kemalpaşazade (d. 1534)[12]. While we know that Kemalpaşazade used earlier sources and relied on an oral tradition based on eye-witnesses of the original events, the problem persists : he might have projected the terminology of his own time.

In view of these obstacles, it is hardly surprising that some European Otto-manists historians — trained originally in either classical philology or medi-eval and early modern European history, and thus well aware of the methods of source criticism — expressed their concerns about the first references to firearms in Ottoman chronicles[13]. Justified these cautious approaches may be, the knowledge of firearms in the 1380s and 1390s, and their more frequent use during the first half of the fifteenth century throughout the Eurasian theater of war indicates that the new weapon made a quick career among the peoples of non-western societies, a fact often overlooked in western narratives of the " gunpowder epic ". Given this overall evolution of gunpowder technology, it is very likely that the Ottomans, who faced enemies already in the possession of firearms, also had knowledge of gunpowder weaponry before the close of the fourteenth century[14].

THE AGE OF OTTOMAN FIREPOWER AND MILITARY SUPERIORITY

The second half of the fifteenth century and the first half of the sixteenth century was a period of Ottoman military superiority. Thanks to Mehmed II's (r. 1451-1481) personal interest in military science and to his patronage, as well as to the frequency of military conflicts in the Balkans and the eastern Mediterranean, Ottoman gunpowder technology succeeded in making up the initial drawbacks of a somewhat belated start within a couple of decades.

The Imperial State Cannon Foundry (*Tophane-i Amire*) in Istanbul, estab-lished by Mehmed II after the capture of the city, was one of (if not *the*) first arsenal in late medieval Europe that was built, operated and financed by a cen-tral government, in a time when most of Europe's monarchs acquired their can-nons from smaller artisan workshops, under the direction of individual gun-founders scattered in cities of the monarchs' respective realms. In the late fif-teenth and early sixteenth centuries, the Imperial Cannon Foundry, Armory (*Cebehane-i Amire*), Gunpowder Works (*Baruthane-i Amire*) and Arsenal *(Ter-*

12. Mücteba Ilgürel, " Osmanlı Topçuluğunun Ilk Devri ", *Hakkı Dursun Yıldız Armağanı*, Ankara,Türk Tarih Kurumu, 1995, 285-293.

13. Vernon J. Parry, " Barud ", *Encyclopedia of Islam,* New edition, London, Leiden, E.J. Brill, 1960, 1061.

14. G. Ágoston, " Ottoman artillery and European military technology in the fifteenth to sev-enteenth centuries ", *Acta Orientalia Academiae Scienciarum Hungaricae*, 47 (1994), 15-48, espe-cially 15-26.

sane-i Amire) gave Istanbul probably the largest military industrial complex in
early modern Europe, rivaled only by Venice. Some of the earliest account
books of the Imperial State Cannon Foundry illustrate Ottoman production
capabilities.

Table 1.

Output of the Imperial State Cannon Foundry in Istanbul, 1513-1528[15]

Date	Length of operation	Number of cannons cast	Material used (metric tons)
1513	4 months	188	27.4
1515-1518	32 months	2 large cast iron	6.3
1517-1518	8 months	24 (of which 22 cast iron)	185
1517-1519	28 months	699 (+428 repaired)	550
1522-1526	38.5 months	1029	483
1527-1528	9 months	148	65
Total	119.5	2090	1316.7

Besides Istanbul, the Ottomans cast cannon in their provincial capitals and
mining centers, as well as in foundries established during campaigns. Of these,
the foundries of Avlonya and Prevesa in the Adriatic (also important naval
bases), Bac, Semendire, Škodra, Praviště and Belgrade in the Balkans ; Buda
and Temesvár in Hungary ; Diyarbekir, Erzurum and Mardin in Asia Minor ;
Baghdad and Basra in Iraq, and Cairo in Egypt were among the important
ones, and were active from time to time. The production output of some of
these foundries could easily match that of the Istanbul foundry. For example,
during the Venetian-Ottoman war of 1499-1503, from October 31, 1499 until
August 26, 1500 the Ottomans cast 288 cannons in Avlonya for the navy. Of
those, 53 were large and medium-size cannons, 29 smaller ones to be used on
board of caiques, and 206 smaller guns called *prangi*[16]. This volume of production was considerable. It is worth noting that it took almost two years for

15. Istanbul, Prime Ministerial Archives (BOA), Kamil Kepeci Tasnifi (KK) 4726. This document was first used by Idris Bostan in his " XVI Yüzyıl Başlarında Tophane-i Amirede Top Döküm
Faaliyetleri " (to be published in the Inalcık Armağanı). The table is based on Bostan's data. For
the years of 1522-1528, see Colin Heywood, " The Activities of the State Cannon-Foundry
(Tophane-i amire) at Istanbul in the early sixteenth century according to an unpublished Turkish
source ", *Prilozi za orijentalnu filologiju*, 30 (1980), which is based on Maliyeden Müdevver
Defterleri (MAD) 7668.
16. BOA, Ali Emiri (AE), Bayezid II, n° 41, and Idris Bostan in his " XVI Yüzyıl Başlarında
Tophane-i Amirede Top Döküm Faaliyetleri ", *op. cit. Prangi* was one of the smallest guns (usually firing shots of 50 *dirhems* or 150 grams) used in large quantities in fortresses and during campaigns. For an attempt at classification of Ottoman artillery pieces, see, G. Ágoston, " Ottoman
artillery and European military technology in the fifteenth to seventeenth centuries ", *op. cit.*

William Lewett to manufacture 120 cast-iron cannons in his foundry in Sussex in 1543-1545. As late as 1679, Seville's foundry, perhaps the most important ordnance factory in seventeenth-century Spain, could hardly manufacture more than 36 cannons of medium caliber per year[17].

Despite the temporary importance of local foundries, Istanbul remained the center of the Ottoman weapons industry. This early centralization of the Ottoman weapons industry was instrumental for the advancement of military technology. Furthermore, this military industrial complex in the capital, supplemented by cannon-foundries, gunpowder workshops and arsenals of smaller scale in the provinces, enabled the Ottomans to establish a clear and long-lasting firepower superiority in Eastern Europe, the Mediterranean, and the Middle East. At Çaldıran (1514) the Ottomans had some 500 cannons, whereas the Safavids had none. At Mohács (1526) the Ottomans employed some 240 to 300 cannons whereas the Hungarians had 85, but only 53 were used during the battle. After the capture of Belgrade in 1521 the Ottomans deployed 200 cannons in the fortress (several of those were certainly of Hungarian origin), and according to an inventory, compiled after the conquest of Rhodes, this fortress had 685 cannons, while the smaller castles of the island had 163 artillery pieces in total[18]. However, the technologically-driven approach should not be taken too far. Even in the most often cited cases in respect of the effectiveness and decisiveness of firearms — the battles of Çaldıran (1514), Marj-i Dabik (1516), Ridaniyya (1517) or Mohács (1526) —, other factors such as numerical superiority, cavalry charge, better logistics and tactics, or profiting from terrain and the mistakes of the enemy, to name but a few, proved to be as crucial as firepower superiority[19].

Although recent research with regard to the size of European armies has shown that the popular figures quoted in the literature are unreliable, it is obvious that the Ottomans had numerical and logistical superiority over their enemies in the period under discussion. The campaign of Mehmed II against Uzun Hasan in 1473 mobilized 100,000 men-at-arms, a body of men which included 64,000 timariot *sipahis*, 12,000 Janissaries, 7,500 cavalry of the Porte, and 20,000 *azabs*. The central imperial budget dated 1528 numbered some 120,000-150,000 members of the regular units, including 38,000 provincial *timar*-holders, 20,000-60,000 men-at-arms brought to the campaigns by the

---

17. C.M. Cipolla, *Guns, Sails and Empires, op. cit.,* 39, 155.

18. I. Bostan, " XVI Yüzyıl Başlarında Tophane-i Amirede Top Döküm Faaliyetleri ", *op. cit.* ; Jenő Gyalókai, " A mohácsi csata " in Imre Lukinich (ed.), *Mohácsi Emlékkönyv*, 1526 (Budapest, 1926), 198, 218.

19. For a more cautious assessment of the importance of technology see, J. Black, *War and the World. Military Power and the Fate of the Continents, 1450-2000*, New Haven, London, Yale University Press, 1998. With regard to the Ottomans, see R. Murphey, *Ottoman Warfare, 1500-1700*, New Brunswick, Rutgers University Press, 1999.

timar-holders, and 47,000 mercenaries (including 24,000 members of the salaried troops of the Porte and 23,000 fortress guards, *martalos* and navy). These figures do not include the various auxiliary troops, who were the following : the *müsellems*, who repaired the roads and bridges in front of the marching armies ; the *yayas*, who helped to transport the cannons ; the *Yörüks*, who collected the draught animals and cast the cannon balls in the mines ; the *cerehors*, who performed various engineering work ; the *akıncı* raiders ; and the Tatar troops[20]. Even the most conservative estimates suggest that at the battle of Mohács Süleyman had at least 60,000 men at his disposal, whereas the total strength of the Hungarian army was only 26,000 men[21].

Ottoman firepower superiority, combined with numerical and logistical superiority, proved to be crucial in mounting a continuous pressure on Europe. To match Ottoman firepower prompted a series of European countermeasures. These included modernization of fortress systems (the introduction of *trace itallienne* into central and eastern Europe) ; changing the cavalry-infantry ratio ; improving the training and tactics of field armies ; increasing the quality and production output of armaments industries ; as well as modernizing state administration and finances. While all these were part of a larger phenomenon, often referred to as the 'European military revolution', (a concept that has recently been under attack) and were undoubtedly fostered by the frequency of interstate violence within Europe[22], in eastern Europe it was Ottoman military superiority, of which firepower was a part, that constituted the greatest challenge and required adequate countermeasures. Thus, in the long run, Ottoman firepower superiority fostered technological experimentation and the increase of industrial production of weaponry in central and eastern Europe, most notably in the Habsburg Empire. In other words, the Ottomans were integral part of the process what might be called the sixteenth-century arms race.

CANNONS AND THE MYTH OF OTTOMAN TECHNOLOGICAL INFERIORITY

Western historians overemphasize the contribution of European technicians to the evolution of the Ottoman gunpowder technology. While it is certainly true that European gun-founders played some role in the advancement of Ottoman casting technology, and provided continuous stimuli throughout the period

---

20. H. Inalcık, " The Ottoman State : Economy and Society, 1300-1600 ", in H. Inalcık and D. Quataert (eds), *An Economic and Social History of the Ottoman Empire, 1300-1914*, Cambridge, Cambridge University Press, 1994, 88-89.

21. Gyula Káldy-Nagy, " Suleimans Angriff auf Europa ", *Acta Orientalia Academiae Scienciarum Hungaricae*, XXVIII, 2 (1974),170-176.

22. G. Parker, *The Military Revolution. Military Innovation and the Rise of the West, 1500-1800, op. cit.* ; J. Black, *A Military Revolution ? Military Change and European Society, 1550-1800*, London, Macmillan, 1991, and *idem*, *European warfare 1494-1660*, London, Routledge, 2002.

under discussion, it should not be forgotten that the majority of the cannon-founders in the Tophane were Turks or other subjects of the sultan. Coming initially from the mining centers of medieval Serbia, Bosnia, Greece and Asia Minor, they brought into Istanbul a diverse and valuable knowledge of metallurgy. This included certain European techniques, such as medieval Saxon mining and metallurgical techniques, known in parts of Serbia and transmitted by Serbian experts to the Ottomans. The other component, perhaps a more important one, was the knowledge of metalworking techniques of the Islamic East that produced the world-famous Damascus blades. It is hardly surprising that sixteenth-and seventeenth century Istanbul, with its Turkish and Persian artisans and smiths ; Bosnian, Serbian, Turkish, Italian, German, and later French, English and Dutch gun-founders and engineers ; as well as with its Venetian, Dalmatian and Greek shipwrights and sailors proved to be an ideal environment for 'technological dialogue'. From the Europeans the Ottomans learned how to make better powder, cannon, mortars, bombs, arquebuses, and ships. The Jews and Marranos, expelled from western Europe at the end of the fifteenth century, supposedly taught them how to construct gun-carriages, an important device responsible for the relative mobility of Ottoman artillery[23].

On the other hand, the Ottomans, of whom one eminent military historian wrote that they were " expert imitators, but poor innovators "[24], are said to have perfected casting technology, and seem to have been instrumental in the introduction of a new lock mechanism, called the serpentine, into Europe. More importantly, they had a long-lasting superiority in making especially strong and reliable musket barrels, using flat sheets of steel, similar to that of the Damascus blades, which were coiled into a spiral, producing great strength in the barrel that could withstand higher explosive pressure. Ottoman musket barrels, soon to be manufactured in Mughal India, were less likely to burst than European barrels with longitudinal seams[25]. Furthermore, the Ottomans played an important role in the introduction and proliferation of firearms in the Khanates in Turkistan, the Crimean Khanate, Abyssinia, Gujerat in India and the Sultanate of Atche in Sumatra. They sent cannon and hand-held firearms to the Mamluk Sultanate, Gujerat, Abyssinia and Yemen (before the latter two were incorporated into the empire), as well as to Mughal India. Even the Safavids, the Ottomans' main rivals in the East, acquired Ottoman artillery and muskets as a consequence of Prince Bayezid's rebellion and escape to Iran. However, it is not clear whether the similarities between, say, Ottoman and Indian techniques for making musket barrels are the results of technological transfer or of

23. V.J. Parry, " Barud ", op. cit., 1062. For the concept of 'technological dialogue' see, Arnold Pacey, Technology in World Civilization, Oxford, Basil Blackwell, 1990, 51.

24. G. Parker, The Military Revolution. Military Innovation and the Rise of the West, 1500-1800, op. cit., 127.

25. A. Pacey, Technology in World Civilization, op. cit., 80-81.

a common technological heritage of Islamic metallurgy. The transfer of know-how from the Ottomans to Safavid Persia and Mughal India seems to be more direct in the case of the diffusion of the *tabur* system, a modified and further developed version of the *Wagenburg* or " wagon fortress ", which was named as such after the Hungarian (*szekér*) *tábor*, *i.e.* " wagon camp ". Even in this instance, however, one should note that the use of carriages against cavalry charges had been known in Central Asia for centuries before the appearance of the Ottoman *tabur*[26].

Recent research has demonstrated that there was an intensive Ottoman-European military acculturation through the common pool of military experts, direct military conflicts and the prohibited trade in weaponry and ammunition during the sixteenth and seventeenth centuries. The employment of foreign experts ensured that new technology and know-how was disseminated relatively fast within and outside Europe. As a consequence, it became virtually impossible to gain any significant technological superiority in the long run. However, echoing Carlo M. Cipolla's almost a half-century-old notion, recent Western historiography claim that the Ottoman artillery hardware missed out on developments in European artillery from the mid-fifteenth century onwards. Whereas in Europe emphasis was increasingly laid on the more mobile light field artillery, the Ottoman ordnance continued to be characterized by giant cannons[27].

Acculturation, however, not only questions the Ottomans' supposed technological inferiority but also suggests close similarities in military hardware. It has been noticed that such similarities are reflected in military terminology : names of the most often used Ottoman cannons, *i.e.*, *balyemez*, *bacaluşka*, *zarbzen*, *kolunburna* or *prangi* are distorted versions of Italian, Spanish, Catalan and Portuguese designations of well-known European firearms and reflect the common material culture of the Mediterranean[28]. Other weapons, such as the small-size *şakaloz*, which got its name from a similarly small Hungarian

26. V.J. Parry, " Barud ", *op. cit.*, 1062 ; H. Inalcık, " The Socio-Political Effects of the Diffusion of Fire-arms in the Middle East ", *War, Technology and Society in the Middle East, op. cit.*, 202-211 ; Salih Özbaran, " The Ottomans' Role in the Diffusion of Fire-arms and Military Technology in Asia and Africa in the 16th Century ", *Revue Internationale d'Histoire Militaire*, 67 (1988), 77-83 ; G. Ágoston, " Muslim-Christian Acculturation : Ottomans and Hungarians from the fifteenth to the seventeenth centuries ", in B. Bennassar and R. Sauzet (eds.), *Chrétiens et Musulmans à la Renaissance*, Paris, Honoré Champion, 1998, 291-301.

27. C.M. Cipolla, *Guns, Sails and Empires, op. cit.*, 95-99 ; G. Parker, *The Military Revolution. Military Innovation and the Rise of the West, 1500-1800, op. cit.*, 126 ; and more recently Jonathan Grant, " Rethinking the Ottoman 'Decline' : Military Technology Diffusion in the Ottoman Empire, Fifteenth to Eighteenth Centuries ", *Journal of World History*, 10, 1 (Spring, 1999), 191-192.

28. On the various types of Ottoman cannons see, G. Ágoston, " Ottoman artillery and European military technology in the fifteenth to seventeenth centuries ", *op. cit.*, 33-45.

hand cannon, called *szakállas* (*puska*) (*i.e.*, a hand cannon or handgun with a beard) reflect the Central-European component of Ottoman weaponry[29]. The *szakállas* (*puska*) is the Hungarian equivalent of the Latin " (*pixis*) *barbata* " or " hook gun " and refers to a large-caliber hand firearm with a hook. The hook served to fix the weapon firmly to the rampart in order to take the firearm's heavy recoil. Hungarian *szakállases* and Ottoman *şakalozes*, which unlike the Hungarian weapons seem to had been set on stock (*kundak*) and transported by gun carriages, composed a considerable part of the artillery in both Hungarian and Ottoman fortresses.

Production output accounts of the Istanbul state cannon foundry as well as inventories of such strategically important Ottoman fortresses as Baghdad, Belgrade and Buda convincingly demonstrate that the overwhelming majority of Ottoman cannons was small and medium-size cannon, and that the Ottomans had a long tradition of smaller artillery pieces. For example, 97 per cent of the 1,027 guns cast at the Imperial State Cannon Foundry in Constantinople in the four years before the battle of Mohács (1526) consisted of small- and medium-sized cannons. 72 per cent of the 300 cannons cast at the foundry in 1685/86 fired cannon balls of 1.28, 0.64 and 0.32 kgs. Contemporary narrative descriptions of the Ottoman weaponry in operation reveal that the use of small-caliber cannon on battlefield was a common practice in the Ottoman army[30]. The available evidence shows that even though Ottoman artillery pieces were heavier than most of the European ones of the same caliber, this seldom constituted serious logistic difficulty for the Ottomans.

In the face of recent research the supposed " metallurgical inferiority " of Ottoman cannon[31], which is based on the superficial assessment of random evidence, also needs re-evaluation. According to chemical analysis, an Ottoman gun-barrel cast in 1464 for Mehmed II was composed of excellent bronze containing 10.15 per cent of tin and 89.58 per cent of copper. The bronze of almost the same composition was recommended by Vanoccio Biringuccio (1480-1539), and was used in contemporary Europe. Ottoman techniques of cannon casting, as described in Michael Kritovoulos's account of casting a

29. This type of weapon was also known to the Rumanians and South-Slavic people who borrowed the name of the gun from the Hungarian and called it *sacalas* and *sakalus* respectively. *Cf.* Lajos Tamás, *Etymologisch-historisches Wörterbuch der Ungarischen Elemente im Rumanische*, Budapest, Akadémiai Kiadó, 1966, 685, and Hadrovics László, *Ungarische Elemente in Serbcroatischen*, Budapest, Akadémiai Kiadó, 1985, 444. *Cf.* also Lajos Fekete, " Az oszmán-török nyelv hódoltságkori magyar jövevényszavai ", *Magyar Nyelv*, XXVI (1930), 264.

30. G. Ágoston, " Ottoman artillery and European military technology in the fifteenth to seventeenth centuries ", *op. cit.*, 43-45.

31. Suggested by G. Parker, *The Military Revolution. Military Innovation and the Rise of the West, 1500-1800*, *op. cit.*, 128.

large cannon for the siege of Constantinople (1453), seems to have been similar to the technology applied in contemporary Europe[32]. Furthermore, Ottoman
production data suggest that, at least until the end of the seventeenth century,
Ottoman cannon founders used the typical tin bronze, which contained 8.6-
11.3 per cent of tin and 89.5-91.4 per cent of copper.

Table 2.

Composition of Ottoman bronze cannons[33]

Date	Copper	Tin
1464	89.58 %	10.15 %
1517-23	91 %	9 %
1522-26	90.5 %	9.5 %
1604	90.8 %	9.2 %
1685/86	91.4 %	8.6 %
1693/94	89.5 %	10.5 %
1704/06	89.6 %	10.4 %
1704/06	89.5 %	10.5 %
1704/06	88.7 %	11.3 %
1706/07	89.5 %	10.5 %

The above data ought to be handled with caution. Despite similar composition sloppy foundry techniques or impurities in the metal might have caused
significant porosity.

As we have seen the Ottomans cast iron guns at the very beginning of the
sixteenth century, that is as early as the English, and a century earlier than their
main sixteenth-century rivals, the Spaniards, who were not capable of producing cast-iron ordnance until the seventeenth century. Furthermore, the first

32. V.J. Parry, " Barud ", op. cit., 1061 ; Jerzy Piaskowski, " The Technology of Gun Casting
in the Army of Muhammad II (Early 15th Century) ", Proceedings of the I[st] International Congress
on the History of Turkish-Islamic Science and Technology, 14-18 September 1981, vol. III, Istanbul, 1981, 163-168.

33. Taken from G. Ágoston, " Osmanlı Imparatorluğu'nda Harp Endüstrisi ve Barut
Teknolojisi (1450-1700) ", in Güler Eren et al. (eds), Osmanlı, vol. 6. Teşkilat, Ankara, 2000, 629,
and based on J. Piaskowski, " The Technology of Gun Casting in the Army of Muhammad II
(Early 15th Century) ", op. cit., 168 (for 1464) ; Istanbul, BOA, MAD 7668 ; C. Heywood, " The
Activities of the State Cannon-Foundry... ", op. cit., and I. Bostan, " XVI Yüzyıl Başlarında
Tophane-i Amirede Top Döküm Faaliyetleri ", op. cit. (for 1517-1523 and 1522-1526) ; MAD 2515
(for 1604) ; MAD 4028 and DBŞM TPH 18597, 18598 (for 1685/86) ; MAD 5432 (for 1693/94) ; MAD
2652 (for 1704-06) and MAD 2679 (for 1706/07). Our calculation should be regarded as approxi-
mate for it is based on the amount of raw material used during casting.

cast-iron guns that were manufactured in England in 1509-1513 were large mortars with detachable chambers and not longer guns cast in one piece[34], which seems to be the case with regard to the Ottoman guns made in Istanbul between 1513 and 1518. Although, the Sultans' master-founders continued to manufacture cast-iron ordnance in smaller quantities in the centuries to come, and the Ottomans, like many Europeans, used them in their navy, bronze cannons remained the dominant weapon in the empire[35]. However, it was a matter of choice, based on the abundance of copper in the empire, and certainly not a question of lack of knowledge of the required technology. Furthermore, it was well-known to contemporaries that bronze cannons, although more expensive than iron ordnance, were safer and less liable to fractures, and less subject to corrosion, especially on board ships.

GUNPOWDER PRODUCTION

Ottoman gunpowder production also shows close similarities to European gunpowder manufacturing. It seems that until the latter part of the seventeenth century Ottoman powder contained 69 per cent of saltpeter, 15.5 percent of sulfur, and the same percentage of charcoal. From the end of the seventeenth century Ottoman powder-works produced powder according to the new mixture (*be ayar-i cedid*) or to the so-called English proportion (*be ayar-i perdaht-i Ingiliz*). Such powder contained 75 per cent of saltpeter and 12.5 per cent of charcoal and sulfur each. This was the most common proportion in England and in most of the European countries even in the first half of the eighteenth century[36]. Though during the second half of the eighteenth century Ottoman gunpowder was still manufactured according to this formula, by the end of the eighteenth century (1794/95) the Ottomans produced a better quality of gunpowder mixed in the proportions of 76-14-10, which closely followed the standard European proportions of 75-15-10[37]. However, maintaining consistent standards was impossible in an empire where gunpowder production was highly decentralized[38]. In the sixteenth century, besides Istanbul, gunpowder was manufactured in Cairo, Baghdad, Aleppo and Yemen in the Arab provinces ; in Buda, Esztergom, Pecs, Temesvár and Belgrade in the European provinces ; as well as in Erzurum, Diyarbekir, Oltu, and Van in Asia Minor. In the seventeenth century the Ottomans established major gunpowder works in

34. H.R. Schubert, *History of the British Iron and Steel Industry*, London, 1957, 249ff ; C.M. Cipolla, *Guns, Sails and Empires, op. cit.,* 39.

35. KK 4726, 18-22, 42-46 ; MAD 2730, 8-10 ; MAD 4688. *Cf.,* also I. Bostan, *Osmanlı Bahriye Teşkilatı : XVII. Yüzyılda Tersane-i Amire*, Ankara, Türk Tarih Kurumu, 1992, 174-77.

36. G. Ágoston, " Gunpowder for the Sultan's Army : New Sources on the Supply of Gunpowder to the Ottoman Army in the Hungarian Campaigns of the Sixteenth and Seventeenth Centuries ", *Turcica*, XXV (1993), 75-96.

37. BOA DBŞM BRI 18321.

38. R. Murphey, *Ottoman Warfare, 1500-1700, op. cit.,* 14.

Bor (in the province of Karaman), Salonica, Gallipoli and Izmir. To be sure, the decentralized nature of gunpowder manufacturing and the resulting inconsistencies with regard to mixture and quality, was not unique to the Ottomans. All those European empires and states that got their powder from similarly varying supply sources faced the same difficulties. It is also important to bear in mind that there was no standardized mixture of gunpowder in Europe. In the mid-sixteenth century, for example, more than 20 different types of powder were produced in Europe, the saltpeter content of which fluctuated between 50 and 85 per cent[39].

Table 3.

Mixture of gunpowder in selected European countries
and in the Ottoman Empire, 1560-1795[40]

Date	Country	Saltpeter	Charcoal	Sulfur
1560	Sweden	66.6 %	16.6 %	16.6 %
1595	Germany	52.2 %	26.1 %	21.7 %
1598	France	75.0 %	12.5 %	12.5 %
1608	Denmark	68.3 %	23.2 %	8.5 %
1649	Germany	69.0 %	16.5 %	14.6 %
1650	France	75.6 %	10.8 %	13.6 %
1673	Ottoman Empire	69.0 %	15.5 %	15.5 %
1686	France	76.0 %	12.0 %	12.0 %
1696	France	75.0 %	12.5 %	12.5 %
1696/97	Ottoman Empire	75.0 %	12.5 %	12.5 %
1697	Sweden	73.0 %	17.0 %	10.0 %
1699/1700	Ottoman Empire	75.0 %	12.5 %	12.5 %
1700	Sweden	75.0 %	9.0 %	16.0 %
1742	England and Europe	75.0 %	12.5 %	12.5 %
1793/94	Ottoman Empire	77.1 %	12.5 %	10.4 %
1794/95	Ottoman Empire	75.8 %	13.7 %	10.5 %

What was more troubling for the Ottomans was the decrease in production output. While in the sixteenth century Ottoman powder-mills could manufacture some 18,000 to 23,000 *kantars* of powder, this amount had fallen to 8,000 to 10,000 *kantars* by the late seventeenth century. In the 1770s and 1780s the

39. G. Ágoston, " Gunpowder for the Sultan's Army... ", *op. cit.*, 78, 87-96.
40. *Ibidem*, and G. Ágoston, " Osmanlı Imparatorluğu'nda Harp Endüstrisi ve Barut Teknolojisi (1450-1700) ", *op. cit.*, 630.

powder-mills at Salonica and Gallipoli were supposed to produce 2,000 *kantars* of gunpowder each ; however, both gunpowder works had serious difficulties in fulfilling these expectations[41]. Consequently, the Porte had to import some 1,500-1,700 *kantars* in certain years from Sweden[42]. However, by the end of the eighteenth century the gunpowder works at Azadlı, which had been modernized by French assistance, were able to manufacture sufficient quantities of gunpowder of a much better quality[43]. In 1800 they supposedly produced 10,000 kantars of gunpowder of good quality[44].

∗∗∗

As a conclusion we may argue that various channels of military acculturation enabled the Ottomans to keep pace with the developments of European military technology well into the seventeenth century. However, by the late seventeenth century European technological superiority of some sort, reported by several contemporaneous observers, must have been a reality. Furthermore, by that time the adversaries of the Porte were able, perhaps for the first time, to match Ottoman firepower and logistics, albeit in terms of deployability of weaponry only in coalition warfare. Yet even this argument should not be taken too far. Ottoman economic and military resurgence in the first half of the eighteenth century, and the success of Ottoman armies against Russia in 1711, and against Austria in the war of 1737-1739 should not be forgotten and needs further research[45]. The real turning point seems to have arrived only by the late eighteenth century, by which time the Europeans outstripped their mighty rival not only in the field of weapons industry, military technology and know-how, but also in such fields as production capacity, finance, bureaucracy, scientific infrastructure and state-patronage[46]. In the long run, all these were of considerable importance for promoting the development of the European military machine, and its subsequent global dominance.

41. BOA, Cevdet Askeriye 9594 and 9595. The standard Ottoman *kantar* weighed 56.449 kgs or 124.43 pounds.

42. BOA, MAD 10398, 102, and MAD 10405, 99.

43. Ekmeleddin Ihsanoğlu, " Ottoman Science in the Classical Period and Early Contacts with European Science and Technology ", in Ekmeleddin Ihsanoğlu (ed.), *Transfer of Modern Science and Technology to the Muslim World*, Istanbul, IRCICA, 1992, 28-29 ; Stanford Shaw, *Between Old and New : the Ottoman Empire Under Sultan Selim III, 1789-1808*, Cambridge, MA., 1971, 143-144.

44. BOA, Cevdet Askeriye 9756.

45. See also, ent of the importance of technology see, J. Black, *War and the World, op. cit.,* 104.

46. On Ottoman science see, Ekmeleddin Ihsanoğlu, *Büyük Cihad'dan Frenk Fodulluğuna*, Istanbul, Iletişim, 1996 ; *idem* (ed.), *History of the Ottoman State, Society and Civilisation*, vol. 2, Istanbul, IRCICA, 2002 ; *idem*, " Osmanlıların Batı'da Gelişen Bazı Teknolojik Yeniliklerden Etkilenmeleri ", *Osmanlılar ve Batı Teknolojisi Yeni Araştırmalar Yeni Görüşler*, in Ekmeleddin Ihsanoğlu (ed.), Istanbul, I.Ü.E.F. Basımevi, 1992, 121-139 ; Aykut Kazancıgil, *Osmanlılarda Bilim ve Teknoloji*, Istanbul, Erkam, 1999.

Un alchimiste vagabond en Europe et au Proche Orient durant le XVIIᵉ siècle

Ioli Vingopoulou

Pareil aux autres ou différent ? Comme les autres ou l'exception ? Voyageur ou savant ? Il avance ou il balance ? Empiriste naïf ou homme de science ? Philosophe-gnostique ou bien un rêveur inquiet ? Oeuvre précursive ou originale ? Singulière ou différenciée du savoir de l'époque ? Oeuvre d'amateur ou bien professionnelle ? Superflue ou fonctionnelle ? Riche ou simplement bavarde ? Peut-être tout cela, peut-être autre que tout ça. Oeuvre émouvante pour les érudits des sciences ou pour ceux qui sont instruits à la recherche du voyage. De toute façon personne ne reste indifférent devant les voyages et l'oeuvre de Balthasar de Monconys[1].

Le voyage est une aventure personnelle ; c'est le besoin de se déplacer, la fuite, la curiosité, la connaissance, l'utilité matérielle, c'est le but à soi ou l'obligation, c'est un besoin ou un " modus vivendi ". Le voyage reste depuis des siècles une expérience personnelle qui accomplit le but à soi mais aussi des besoins et des obligations à travers un processus de transitions du " familier et sûr " à " l'inusité-attrayant mais étrange et risqué ".

---

1. Balthasar de Monconys, *Journal des Voyages de Monsieur De Monconys, Conseiller du Roy en ses Conseils d'Estat et Privé, et Lieutennant Criminel au Siege Presidial De Lyon. Où les Sçavants trouveront un nombre infini de nouveautez, en Machines de Mathematique, Experiences de Physique, Raisonnemens de la belle Philosophie, curiositez de Chymie, et conversations des Illustres de ce siecle ; outre la description de divers Animaux et Plantes rares, plusieurs Secrets inconnus pour le Plaisir et la Santé, les Ouvrages des Peintres fameaux, les Coûtumes et Moeurs des Nations, et qu'il y a de plus digne de la connoissance d'un honeste Hommes dans les trois Parties du Monde. Enrichi de quantité de Figures en taille-douce des lieux et des choses principales, Avec des Indices tres-exacts et tres commodes pour l'usage.Publié par le Sieur de Liergues son Fils*, A Lyon, Chez Horace Boissat, et George Remeus, 1665-1666, t. I-II-III ; suivant : Monconys, *Journal*.

Fils du Lieutenant criminel de Lyon, Balthasar de Monconys naquit en 1611[2]. Pour éviter l'horrible peste qui ravagea cette ville en 1618, ses parents l'envoyèrent faire ses études à Salamanque (1628)[3]. Mais il croit toujours qu'il y a peu de chose à voir en ce pays. Il rêve d'aller jusqu'aux Indes et la Chine mais l'amour que son père avait pour lui et la tendresse d'une belle soeur — la plus vertueuse ! — le font revenir à Lyon.

" De ses commencements il avoit une forte passion pour la Chymie qu'il estimait la clée de la Nature qu'elles nous ouvre par la dissolution des corps. Il chercha par tout ce que personne n'a encore trouvé. Il a voulut tout examiner, et n'oublia rien pour entrer en la connaissance des arts les plus occultes, et les plus abstrus et comme les Esprits extroardinaire peuvent mal-aisement se tenir dans les bornes de la mediocrité... il fut curieux jusqu'à l'excés "[4]. À l'age de trente-quatre ans le lyonnais Conseiller royal abandonne la sûreté d'une bibliothèque et la spéculation approfondie, abandonne l'étude personnelle et la pensée orientée du cabinet et se voue à la recherche, enrichissant ses bagages spirituels, procédant intrépide vers l'Orient et l'Occident[5].

Le Portugal d'abord, en 1645, ou, d'une part la cour admira la promptitude, et la facilité avec laquelle il dressait les Horoscopes et nous, de notre part, nous commençons à découvrir à travers un style d'écriture pédant et plat, sans élan et couleur ce que nous constaterons pour tous ses voyages : un plexus surprenant dans chaque ville avec des " connaissances " (mathématiciens, pères, orfèvres, chirurgiens, machinistes, chimistes, docteurs ou princes, curieux des sciences, philosophes, savants) et ses visites dans des cabinets d'où il commence à enregistrer *secrets et expériences*. Suit son premier voyage en Italie (1646), où à part ses intérêts particuliers, son admiration couvre totalement les arts explosifs et resplendissants de ce pays[6].

Mais son goût de la philosophie l'entraîna à faire le voyage en Orient pour y étudier les différents dogmes professés dans cette partie du globe, y chercher des traces des anciennes religions, des sectes gymnosophistes, astrolâtres etc. Peu après (1647-1648) alors, le voilà en Egypte[7] et au Levant " il battit à la

2. *Biographie Générale Nouvelle*, Paris 1801/Copenague 1968, col. 952.s.

3. Ce voyage est décrit dans le troisième volume : Monconys, *Journal*, vol. III, 1-60.

4. Monconys, *Journal*, vol. I, 2-3.

5. Pour le voyage des savants des époques précédentes ainsi que pour le XVIIe siècle voir l'article de Efthymios Nicolaïdis, " Voyages et Voyageurs ", in M. Blay, R. Halleux (éds), *La Science Classique, XVIe-XVIIIe s., Dictionnaire critique*, Flammarion, Paris 1998, 165-177 ; d'où aussi une orientation bibliograhique sur ce sujet.

6. Ses voyages d'Italie lui procurent l'avantage d'y visiter tout ce qui y reste des disciples du grand Galilée. Il y a connu les : PP. Zucchi, Fabri, Richeome, Kirker, les Sieurs Torriccelli, Viviani, Belluci, Del Pozzo, Divini, Cassino ; Monconys, *Journal*, vol. I, 3-4.

7. Pour son voyage en Egypte voir Balthasar de Monconys, *Voyage en Egypte de Balthasar de Monconys 1646-1647*, Présentation et notes d'Henry Amer, IFAO, Caire, 1973 (Collection des Voyageurs Occidentaux en Egypte, n° 8).

porte de tous ceux que la superstition a érigé en Sages, et s'informa diligement... s'il rencontait parmy eux quel que reste de l'ancienne Cabale des Egyptiens, ou de la Philosophie de Mercure Trismégiste[8] et des Zoroastres, que Pythagore et Platon piquez d'un pareille ardeur ... "[9]. Mais, au lieu de cela on retrouve l'Européen de l'époque à l'Orient. Perdu dans les ruelles des villes et le labyrinthe de ses pensées, fasciné de tout, il nous offre une description excellente de son cheminement, respirant tout le temps les odeurs et observant les moeurs ; se mêle avec les Turcs, les Coptes, les Juifs, les Arabes, les Mores. Mais le texte diffère : aucun secret, aucun remède n'a été rédigé ou plutôt classé parmi les notes du voyage. Il obéit sans protester à des exhortations superstitieuses, il découvre de merveilleux livres d'astronomie mais en hébreu, persien ou arabe. Ni les Maronites, ni les Arméniens, ni les Dervish ou les Arabes auxquels pose des questions ne sont capables de lui répondre[10]. Il a besoin de noter, de rédiger des secrets mais l'Orient est plus anéantissant que sa raison rationnelle. La dualité de sa personnalité se découvre : il cherche, il ne trouve pas, or il rêve et doute de tout. Il veut partir. Il traverse l'Asie Mineure, il arrive à Constantinople (May 1648).

En ce lieu c'est le vrai voyageur du XVIIe siècle[11]. Pérégrinations, admirations, marchés, curiosités et " le plus bel objet en cette ville : la vue "[12]. Mais là encore : " un Turc curieux en Astronomie vint à ma chambre où nos truchements ne pouvoient nous donner la satisfaction que nous souhaitons l'un de l'autre "[13]. Il cherche affolement d'aller plus loin : partir en caravane en Perse. Encore une fois la peste le fait fuir. Avant dernière étape en Orient, à Smyrne

---

8. Comme il advint avec tous les textes de l'antiquité de même pour les textes hermétique les premières tentatives de leur publication se réalisèrent pendant la Renaissance. En 1471 nous avons la traduction en latin des premiers quatorze discours philosophiques hermétiques. D'autres s'en suivent jusqu'à la fin du XVIe siècles. L'hermétisme pratique — d'application — déploya un grand intérêt pour les sciences expérimentales et physiques comme : la Médicine, la Botanique et la Chimie, toujours pourtant sous l'influence de l'Astrologie-Occultisme. L'influence de l'hermétisme sur l'Alchimie — extraits avec des conseils d'Hermès pour la fabrication de l'or — et sur les tendances apocryphes pour les soins médicaux furent déterminantes surtout pendant le bas Moyen Age : *Hermès Trismégiste, Oeuvres Complètes*, t. I, Ed. KAKTOΣ, 2001, 19, 21-23, 25 ; t. IV, 814-817.

9. Monconys, *Journal*, vol. I, 3.

10. Désillusionné plusieurs fois, il cite ses efforts de connaître les savants de ces pays mais sans succès : " Puis je fus à Boulak chercher un More qu'on dit qui apprend les Langues en huit jours, mais je ne le trouvay point ; je fus parler à un autre qui m'avoit promis de me venir trouver le Vendrendy passé, et me faire voir quelque chose extaordinaire ; il me dit qu'il ne scavoit rien... ", Monconys, *Journal*, vol. I, 274.

11. Le voyageur du XVIIe siècle et surtout son récit se caractérisent principalement par le but qui vise à la communication de la connaissance acquise en tant que résultat d'une expérience personnelle qui mêle l'objectivité et la visée générale du message descriptive, la plupart des fois, et la subjectivité des mémoires personnels ; Rania Polycandrioti, " La préface du récit de voyage : les éditions françaises au XVIIe siècle ", *Tetradia Ergasias 17, On Travel Literature and Related Subjects-References and Approaches*, Athènes, Institute of Neohellenic Research, 1993, 539-540.

12. Monconys, *Journal*, vol. I, 389.

13. Monconys, *Journal*, vol. I, 393.

" je trouvay un médecin Turc jadis Chretien, et puis Juif : qui est savant en Médecine, Philosophie, et Mathématique, et qui a cherché autant que moy sans rien trouver "[14]. Or, il revient à Lyon en 1649.

Mais enfin, n'ayant pu rassasier son esprit d'une nourriture si creuse, et si légère ; il tourna toutes ses pensées " à la belle Physique et aux Mathématiques qui de toutes les sciences humaines, sont celles où il y a plus de profit, et de solidité "[15]. Il tourna ses intérêts de recherche en de pays communs où il eut l'amitié et la familiarité de tous les vertueux d'Europe. À partir de 1663-1665 entre Paris et l'Angleterre, la Hollande et l'Allemagne — théâtres où ses rares qualités l'ont le plus attiré — il se retrouve plus dépaysé, dans les milieux des Académies, ou des Palais des Princes, dans les Bibliothèques ou les Cabinets[16]. Et il se fit connaître à une majorité de lettrés et hommes des sciences dans ces pays et le grand nombre des lettres prouve sa réputation[17].

Mais le corps change même si l'âme ressent les mêmes inclinations. Diverses attaques de l'asthme et ses autres maux ne lui permettent pas d'instituer une Assemblée de Physique à Lyon. Conservant toujours dans son coeur un respect infini pour la Divinité, il envisagea la mort avec une intrépidité de Philosophe, mais il s'y disposa avec la crainte et l'humilité d'un Chrétien[18].

Le fruit de ses observations — seulement celles relatives à son journal de bord, et pas les secrets et les expériences — mises en ordre par son fils, Sieur de Liergues, et son ami le savant Jésuite Jean Berthet et publiées en 1665-1666 en trois volumes [planche 1], dépassent les 1300 pages et sont ornés de figures assemblées par planches[19]. Son style est lourd, monotone et diffus. Ses dessins et esquisses naïfs et enfantins [planche 2][20]. Ses remarques curieuses et d'un intérêt global : *Secrets pour les boutons de visages, Pour le mal des dents, la gangrène, la dysenterie, Méditation sur la production naturelle des métaux, et la façon artificielle de la pierre philosophale, secret pour alléger l'or, devinez*

14. Monconys, *Journal*, vol. I, 424. De Smyrne il fait une visite à Chios et il nous laisse parmi les descriptions le plus intéressantes de cette île datant du XVIIe siècle.

15. Monconys, *Journal*, vol. I, 5.

16. Depuis son voyage en Angleterre et les Bays-Bas il tiendra correspondance avec MM. Hobbes, Digby, Boile, Morey, Oldenbourg, Bronker, Vallis, Vrenne et Messieurs Vosius et Sluse et d'autres d'Hollande. Monconys, *Journal*, vol. I, 3

17. À Paris il se fit connaitre à M.M. Gassendi, Bourdelot, Thevenot, Justel, Petit, Roberval, Pascal, De la Chambre, Sorbière, Miramont, Lantin, Henri, Rool, Anzout, Monconys, *Journal*, vol. I, 4.

18. Monconys, *Journal*, vol. I, 7-8.

19. Une seconde édition a paru à Paris et Lyon en 1667. Ils ont suivis d'autres ainsi que une traduction allemamde en 1697, Leonora Navari, *The Blackmer Collection*, London 1989, 241.

20. " Je pris à mon service un peintre Hollandais, afin de joindre la representation des choses que ie verrois à la simple relation, dont il faudra que le lecteur se contetnte... il m'abandonna... " sans preciser si à la suite c'était lui-même ou quelqu'un d'autre qui a fait les esquisses parmi les notes de son *Journal*. Monconys, *Journal*, vol. I, 2.

deux cartes, lampe qui se fournit de l'huile, et de mesche [planche 3][21], *Pour écrire sans qu'il se connoisse, pour la fièvre, observations sur l'éclipse, loupes, Secrets pour les hémoroïdes, Microscopes, thermomètres, expérience de la solution du sel, machine hydrolique, problème d'algèbre, pour peser les liqueurs, etc.*

Découvrons un peu ce matériel : Ses notes de voyage sont chronologiquement rangées mais interrompues par des " secrets et expériences ". Entre les discussions faites à Lisbonne sur l'âge du monde et les opinions de Galilée[22] et ses discussions à Marseille sur les loupes, [planche 4] les volcans, les éclipses et les curiosités d'un cabinet à Pise avec l'instrument pour reconnaître la sècheresse où l'humidité, il est le premier en France à porter des lunettes d'Eustatio Divini. Il rédige plusieurs discours sur le sujet : *connaissances nécessaires pour la constructions des lunettes*[23] sans omettre que " M. Baptista Riccardi qui a beaucoup travaillé à la Chimie il me donna plusieurs secrets "[24] et sa récolte, de son premier voyage en Italie, s'accomplit, entre autres, avec le *Secret pour rendre le poids à l'or*[25] et *La conversion de l'air en eau*[26].

Toujours en contact avec les Consuls ou les Pères installés dans les pays d'Orient, suivi d'un janissaire — comme il fallait durant ses périples — il trouve pourtant des heures ou jours de solitude pour expérimenter sur la *Solution des Sels*[27] ou la *Démonstration de la vélocité acquise des plans inclinés* [planche 5] ou travailler sur l'*Instrument d'Hydrotechnie* [planche 6] et la *Manière de tirer l'huile de soufre*[28]. On suit ses excellentes descriptions de son cheminement au Mont Sinaï et dans le désert [planche 7], [planche 8], [planche 9] (= machine pour tirer l'eau, rompre et couper la paille)[29] et son pélerinage aux Lieux Saints [planche 10] [planche 11] en éprouvant les soirs ses *lunettes à la lune*, retirant par les Juifs ou les Frères les *remèdes pour détacher les Sangsuës*, des observations sur le *Flux de Mer*, la *teinture du cuivre* tout en

21. Monconys, *Journal*, vol. I, 65-68.
22. Entre autres : " Je mis net pour le Père Elzear cette démonstration que l'on m'avoit donnée à Florence *Dimostrazione trovata dal gran Galileo, l'anno 6139* [figure 20]. Il s'agit de la *Démonstration de la vélocité acquise dans les plans inclinés...* Monconys, *Journal*, vol. I, 169-172. En ce qui concerne la révolution dans la pensée scientifique pendant la fin du XVIe et le XVIIe siècle et l'évolution de l'esprit avec l'admission des méthodes expérimentales et mathématiques dans tous les domaines de la Science, qui s'imposa essentiellement d'après l'oeuvre de Galilée voir : A.C. Crombie, *Augustine to Galileo*. Volume II. *Science in the Later Middle Ages and Early Modern Times, 13th to 17th centuries*, London 1979, [traduction grecque : MIET, Athènes, 1992, 125-167.
23. Monconys, *Journal*, vol. I, 117.
24. Monconys, *Journal*, vol. I, 132.
25. Monconys, *Journal*, vol. I, 138.
26. Monconys, *Journal*, vol. I, 134.
27. Monconys, *Journal*, vol. I, 166.
28. Monconys, *Journal*, vol. I, 172, 173, 174.
29. Monconys, *Journal*, vol. I, 266.

nous laissant des dessins indéchiffrables et non rangés, ni ultérieurement [planche 12].

À Constantinople, on lui apporte des remèdes du Sérail, il passe des matinées *à mesurer et peser l'eau* et pour la première fois, il précise que le soir *J'escrivis ces secrets* : suivent six pages de recettes entre autre bien détaillés : *Horloge chimique, paste qui blachit le mercure en le frottant*[30], *guérir un chancre, les cataractes dans l'oeil*, etc. Sa fascination couvre l'architecture, le passé brillant en cette ville, le paysage, les raretés [planche 13] mais finalement comme il cite : " Je rompis mes desseins et fut tout le jour inutile et mélancholique "[31]. Les phénomènes naturels l'attirent lors de son voyage du retour [planche 14][32].

Finalement quatorze ans plus tard il se trouve dans le milieu et l'air pompeux des assemblées de l'Académie de Londres *pour une infinité d'expériences, sur lesquelles* [ils] *ne raissonent point*[33] : mais un fleuve des nouveautés et observations suit sur : les *maschines hydroliques* [planche 15][34], *microscopes, les larmes de verre*[35], *l'élasticité de l'air* ; *les fourneaux*[36], *l'ascension de l'eau, ou curiosités sur les couleurs*, [planche 16], *la pesanteur de l'air, des machines du temps, la dissolution de l'eau, les thermomètres*[37], *balances, instruments à dessinés ou pour la mesure du chaud et du froid, machines pour fournir l'air* [planche 17] et cette richesse surabondante se complète par *le modelle de maschine pour prendre la distance de deux étoiles*[38], *ou maschine par le moyen de laquelle on peut descendre au fond de la mer*[39], *maschine pour faire des experiences du vuide (17-18)* [planche 18] *et la dissolution de l'or* ou *le vaisseau infernal*[40], *la manière de relever en perspective tous le plans*, et *les mines pour l'extraction du vitriol*[41][planche 19].

Au Pays Bas il cite des curiosités architecturales, des problèmes hygiéniques, *la pompe pour jeter l'eau, la manière de préparer l'antimoine ou le savon*[42] [planche 20], [planche 21] et puisque les éditeurs ne trouvent pas dans

30. Monconys, *Journal*, vol. I, 398, 400
31. C'était le 6/11/1648. Monconys, *Journal*, vol. I, 448.
32. Monconys, *Journal*, vol. I, 490.
33. Monconys, *Journal*, vol. II, 25.
34. Monconys, *Journal*, vol. II, 26-29.
35. Monconys, *Journal*, vol. II, 32.
36. Monconys, *Journal*, vol. II, 29, 42.
37. Monconys, *Journal*, vol. II, 54.
38. Monconys, *Journal*, vol. II, 73.
39. Monconys, *Journal*, vol. II, 40.
40. Monconys, *Journal*, vol. II, 76
41. Monconys, *Journal*, vol. II, 79, 81.
42. Monconys, *Journal*, vol. II, 112-113, 116.

les manuscrits des raretés dont l'auteur fait mention ils intercalent des frag-
ments des lettres écrites ou envoyées par/ou pour M. De Monconys[43]. Les
Bibliothèques[44], et les nouveaux livres, les cabinets d'Anatomie[45] et les *mer-
veilles chimiques débitées par M. Borri* ainsi que des entretiens avec lui[46],
ainsi que les opinions de M. Vossius pour *l'Arquebuse* et autres[47] [planche 22]
complètent son voyage en Hollande.

De sa longue déambulation en Allemagne et de son deuxième voyage en Italie voici [planche 23] les *machines à élever l'eau*[48], *le vase pneumatique pour
démontrer la force élastique de l'air*[49] et une infinité d'expériences décrites en
détail[50] [planche 24] ainsi que huit pages de *conseils et enseignements pour les
maux du corps*[51] mis au net avec encore huit pages des *secrets chimiques*[52]
enrichissent la fin du deuxième volume.

Quant au troisième volume — qui ouvre avec son premier voyage en Espa-
gne fait en 1628 à l'age de 17 ans[53] — on se trouve face à un admirable
trésor : il s'agit d'un essai avec des exemples et opérations sur *L'usage et le
pesage des liqueurs*[54] [planche 25], la *Règle pour le Toisage et des expériences
sur la communication des vitesses suivant les doctrines de Galilée*[55] [planche
26], d'un *Traité d'Algèbre et de Géométrie*[56] [planche 27], [planche 28]. Surprise charmante à la fin, ses poèmes : sonnets, épîtres, lettres burlesques, élo-
ges, chansons, épigrammes, et comme par exemple : *Bouts rimés à la bataille
de Marathon !* et plusieurs lettres écrites ou reçues par lui même[57]. Et enfin
une incroyable liste qui couvre trente cinq pages, dont finalement nous ignorons la source de provenance, de cent soixante cinq raisonnements et expériences physiques, médicinales, chimiques et autres non numérotés et classés qui
couvre l'espace des intérêts de cet grand homme[58].

43. Monconys, *Journal*, vol. II, 137-140, 163-168.
44. Comme celle de M. Zulcon, Monconys, *Journal*, vol. II, 149.
45. Monconys, *Journal*, vol. II, 150, 169.
46. Monconys, *Journal*, vol. II, 134-136, 137, 145, 146, 147, 148.
47. Monconys, *Journal*, vol. II, 152-154, 179-180.
48. Monconys, *Journal*, vol. II, 209.
49. Monconys, *Journal*, vol. II, 231-232.
50. Monconys, *Journal*, vol. II, 247-248, 278-279, 341, 342, 347.
51. Monconys, *Journal*, vol. II, 328-335.
52. Monconys, *Journal*, vol. II, 335-341
53. Commencé le 15 septembre ayant comme fidèle conducteur le Sieur Randon et redigé en
1648, Monconys, *Journal*, vol. III, 1-60.
54. Monconys, *Journal*, vol. III, 3-14 (deuxième numérotation du volume)
55. Monconys, *Journal*, vol. III, 15-52.
56. Monconys, *Journal*, vol. III, 1-44 (numérotation apart).
57. Monconys, *Journal*, vol. III, 1-32 (troisième différente numérotation)
58. Monconys, *Journal*, vol. III, 60-96.

Un premier classement — en gros — du matériel est donné dans l'édition avec les index et les tables[59]. En même temps une très longue liste des noms des personnes — spécialistes, savants, docteurs, alchimistes, astrologues, mathématiciens, empiristes — comme lui ou plus spécialisés extraite et établie par mon étude (peut être plus de cinq cent noms) et en combinaison avec la correspondance qu'il tenait pendant ses voyages tresse une intéressante toile sur laquelle les spécialistes des sciences peuvent broder des villes, des hommes et des cabinets pour une période de plus de vingt ans au milieu du XVIIᵉ siècle.

Notre homme réunit des recettes médicinales, des formules chimiques, du matériel pour les pratiques des sciences occultes, des énigmes mathématiques, des problèmes d'algèbre, des observations zoologiques et des applications mécaniques. Il notait, tout, sans arrêt. Mais tout ce matériel nous a été fourni par ses descendants. Il aurait pu, peut-être, aboutir à quelques conclusions s'il l'avait élaboré lui même. Il est tout aussi possible qu'il enlèverait plusieurs renseignements ou bien, se pliant aux exigences et à la stratégie éditoriale de son temps, qu'il altérerait l'ensemble. C'est ainsi que nous sommes — oui — en possession de l'ensemble de ses écrits, mais nous sommes appelés à le valoriser.

Avec modestie, j'ose exprimer une opinion personnelle, pourtant non spécialisée à ces sciences mais sensible au comportement et à la mentalité de voyageur. Je suis probablement en mesure d'extraire et de deviner son désappointement de la récolte de son voyage en Orient, contrairement à son enthousiasme passionné et sa communication avec les personnes et les sujets surtout en Angleterre tout aussi aux Pays-Bas et en Italie ; et un peu moins en Allemagne et au Portugal. L'Orient (l'Egypte, la Syrie, les Lieux Saints, l'Asie Mineure, Constantinople) le confondait plus qu'il ne l'enrichissait, l'attirait, comme il arrivait à tous ceux qui étaient habitués au *modus vivedi* de l'Occident. Toutefois l'Orient le désappointa en ce qui concerne l'assemblage du matériel. Nous, plus flexible et perspicace que sa pensée, nous pourrions en trouver — discerner la cause de ce conflit. La langue, pour les intrépides des longs voyages, c'était le problème primordial. La langue, écrite ou orale, qui contient toute la force intellectuelle d'un peuple, rapprochera ou éloignera les mondes qui cherchent la vérité, qu'elle soit amère ou savante.

Cette désillusion de son voyage en Orient — qui transparait plus qu'elle n'est témoignée dans le texte de Monconys — s'exprime mieux par un autre

---

59. Les Tables Générales de tout ce qui est contenu en ces trois Volumes sont les suivantes : *Table des Villes, et lieux Principaux décrits en ces Voyages ; Table des Hommes Illustres, et Bons Ouvriers, dont il a fait mention en ces trois Volumes ; Table des Machines et Artifices Divers ; Table des Experiences, Observations, et Raisonnements Physiques ; Tables des Animaux, Plantes, Pierres, Oiseaux, et autres raretés de Nature ; Table des Secrets et Receptes et Table des Matières.*

grand voyageur de la fin du XVIIe siècle, Jean Chardin, voyageur vers le royaume de Perse qui pourtant maîtrisait parfaitement le turc et la langue du pays. Il définit : " la Chymie est ordinairement divisée en deux parties l'une destinée à préparer les Remèdes du Corps, l'autre à chercher la pierre Philosophale. A l'égard de la première, les Persans ne connoissent point des Remèdes Chymiques, et ne donnent pas leur médicaments en forme de pilules, ni de Poudres, et quand nous leurs de la quantité de leurs Emulsions et de leurs Potions, qu'ils donnent à pleines terrines que nous leur opposons notre méthode, ils disent que notre climat est différent du leur, et que chaque Pais à ses manières. Pour qui est de l'autre partie de la Chymie les Persans la connaissent comme nous, et ils en font oeuvre plus enfantées, mais la plupart s'y ruinent en Perse, aussi bien qu'on font en Europe et on peut dire qu'ils n'y reussissent pas mieux que nous "[60].

Remarques qu'on pourrait sûrement imaginer un demi siècle auparavant dans le texte de Monconys, lors de sa visite au Proche Orient, s'il arrivait de publier lui même sa relation.

Précoce pour la propagation des sciences en Orient, dans l'esprit de son temps pour l'Occident, il combine le charme d'un journal de bord d'un voyageur, et l'intérêt d'un cahier de notes et remarques d'un érudit de cabinet.

" Ceux qui achètent les livres de Voyages n'ont pas toujours la même visée. Les uns les veulent estudier dans la chambre ; les autres les demandent comme de fideles guides de leur Pèregrinations "[61]. Ici le " Journal des Voyages… " d'un alchimiste vagabond est un amas d'une infinité des choses rares et recherchées.

60. Jean Chardin, *Voyage du Chevalier Chardin en Perse, et autres lieux d'Orient, enrichis d'un grand nombre de belles figures en taille-douce, qui représentent les antiquités et les choses remarquables du païs. Nouvelle édition, augmentée d'un grand nombre des passages tirés du manuscrit de l'auteur, qui ne se trouvent point dans les éditions précédentes*, A Amsterdam, 1735, 274-275.

61. Monconys, *Journal*, vol. I, 9.

IOVRNAL
DES·VOYAGES
· DE MONSIEVR
DE MONCONYS,
Conseiller du Roy en ses Conseils d'Estat & Priüé,
& Lieutenant Criminel au·Siege Presidial de Lyon.

Où les Sçauants trouueront vn nombre infini de nouueautez,
en Machines de Mathematique, Experiences Physiques,
Raisonnemens de la belle Philosophie, curiositez, de Chymie,
& conuersations des Illustres de ce Siecle;

Outre la description de diuers Animaux & Plantes rares, plusieurs
Secrets inconnus pour le Plaisir & la Santé, les Ouurages des Peintres
fameux, les Coûtumes & Mœurs des Nations, & ce qu'il y a de plus
digne de la connoissance d'vn honeste Homme dans les trois Parties
du Monde.

Enrichi de quantité de Figures en Taille-douce des lieux & des choses principales,

Auec des Indices tres-exacts & tres-commodes pour l'vsage.

Publié par le Sieur de LIERGVES *son Fils.*

PREMIERE PARTIE.
Voyage de Portugal, Prouence, Italie, Egypte, Syrie, Constantinople, & Natolie.

A LYON,
Chez HORACE BOISSAT, & GEORGE REMEVS.

M. DC. LIV.
AVEC PRIVILEGE DV ROY.

PLANCHE 2

PLANCHE 3

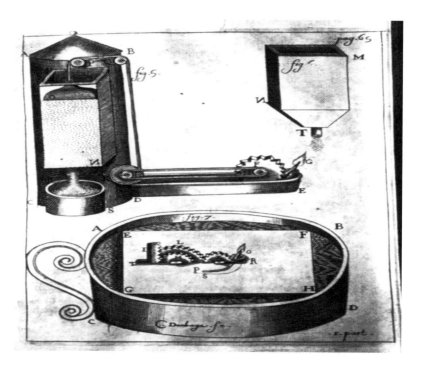

PLANCHE 4

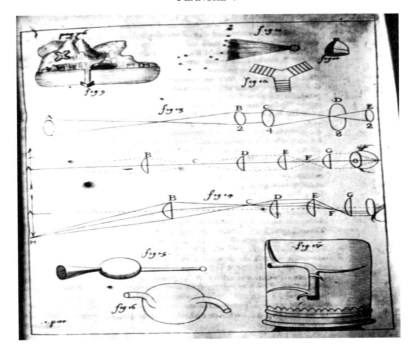

PLANCHE 5

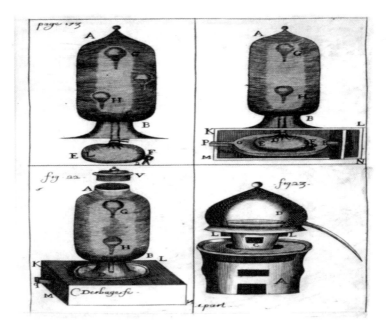

PLANCHE 6

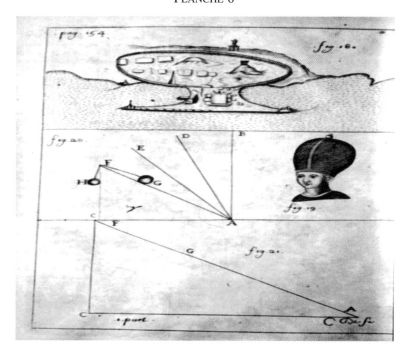

PLANCHE 7

PLANCHE 8

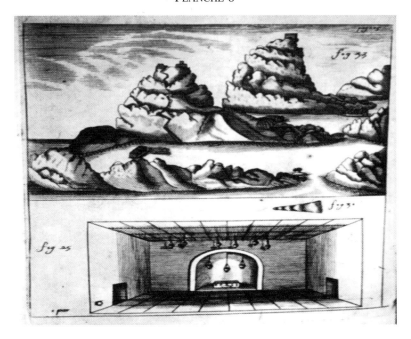

PLANCHE 9

PLANCHE 10

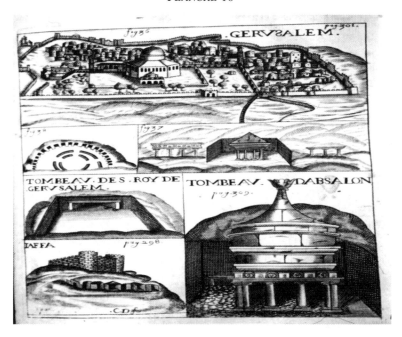

PLANCHE 11

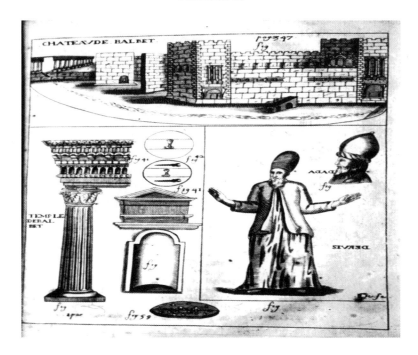

PLANCHE 12

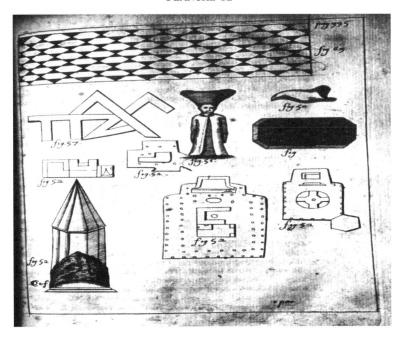

PLANCHE 13

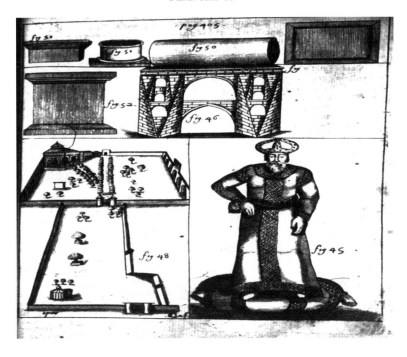

PLANCHE 14

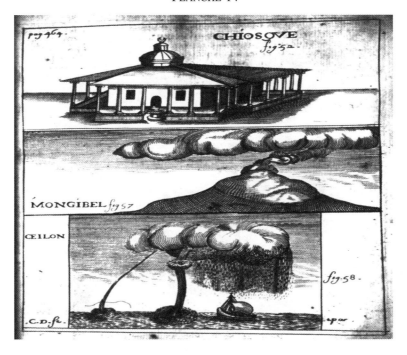

PLANCHE 15

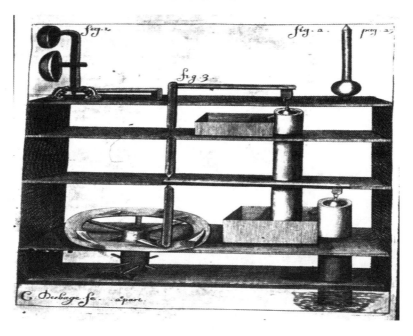

PLANCHE 16

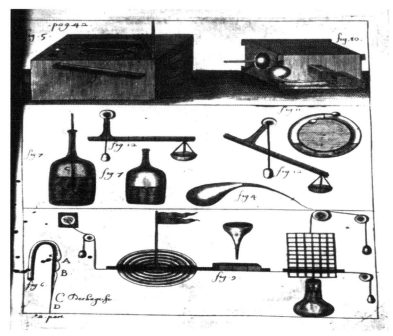

PLANCHE 17

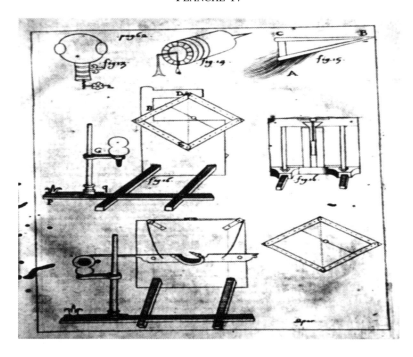

PLANCHE 18

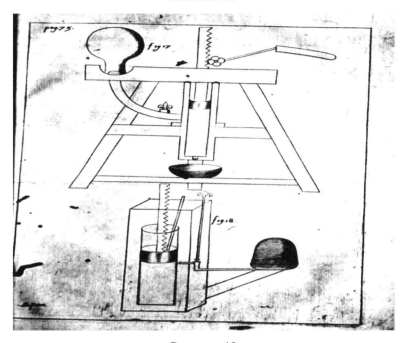

PLANCHE 19

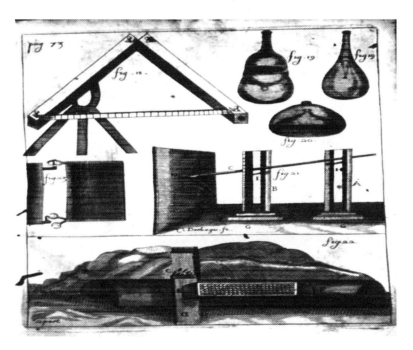

PLANCHE 20

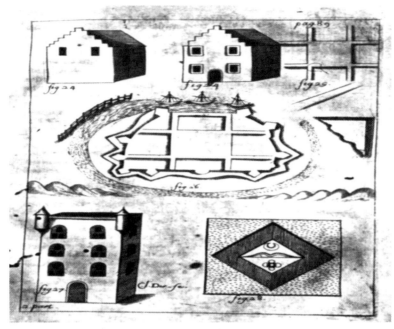

PLANCHE 21

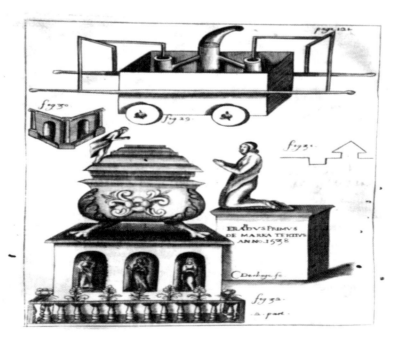

PLANCHE 22

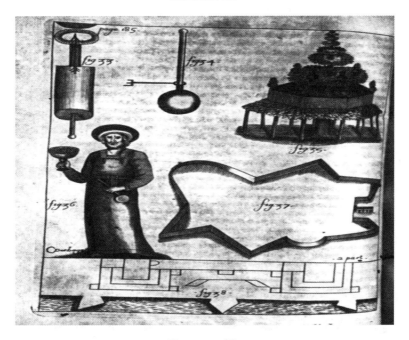

PLANCHE 23

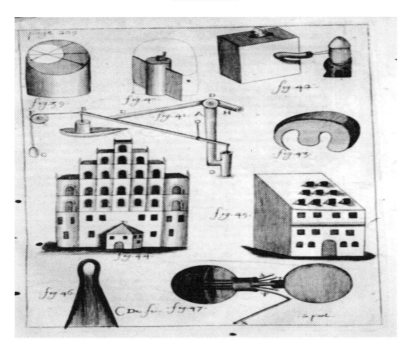

PLANCHE 24

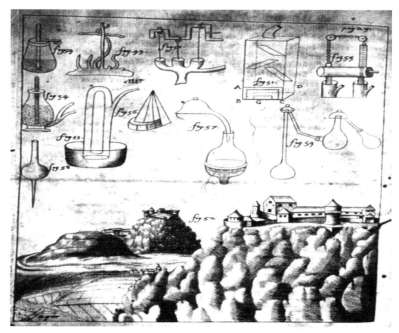

PLANCHE 25

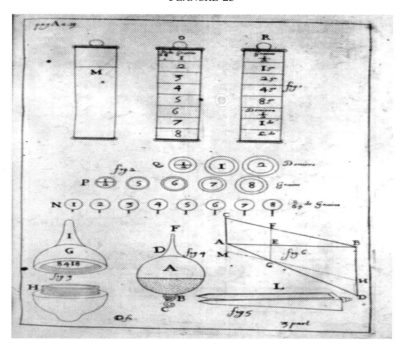

PLANCHE 26

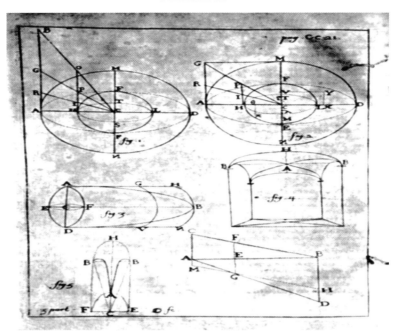

PLANCHE 27

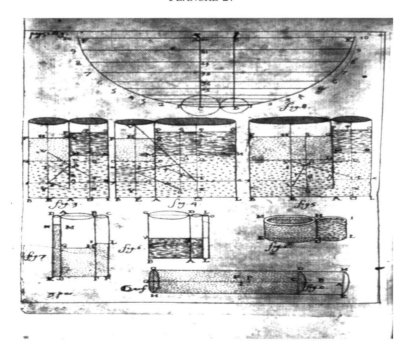

PLANCHE 28

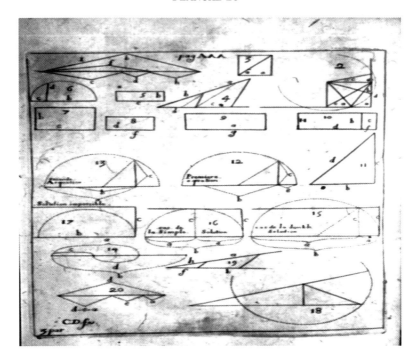

The Confrontation of mathematics on behalf of the Eastern Orthodox Church during the Ottoman period

Maria Terdimou

Research into the confrontation of mathematics by the Church is part of the topic " Church - exact sciences ", which, in turn, is included in the broader framework of the study of the relationship between Eastern Orthodoxy and the Enlightenment movement, particularly in the period from 1750 to 1821.

As known, a gradual differentiation occurred in Greek society during the 18th century and new elements and perceptions were introduced in the existing cultural and social structures. Those new conditions produced the Neohellenic Enlightenment movement. In its effort to lead Greek society to new grounds, the movement came into conflict with the Church and questioned, albeit not always directly, the hitherto dominant role of the Church in social, national and educational affairs. As it was the case in Europe, the most spectacular aspect of the movement was perhaps its anti-religious polemics. In the political, intellectual and ethical conditions of the Old Order, the battle for Enlightenment had to be fought against the Church, although the battle range was certainly much broader.

The dominant view is that as long as mathematics is not confronted as the science which can understand and interpret the world, as was believed e.g., by the Pythagoreans, it can stay clear of any social or religious dispute. All issues relevant to the confrontation of the universe, the existence — or not — of eternity and the meaning of the world are the subject matter of physics, chemistry and astronomy. In this case, mathematics serves solely as a tool necessary in the natural sciences. The subject matter of the natural sciences may be easily popularized and can therefore be easily accessed by the broader reading public. The situation is different for mathematics, as its simplified presentation is not easily feasible.

At first, the above seem to justify satisfactorily the claim that mathematics can stay clear of the fires of religious dispute. In the course of this essay, it will be established whether this is truly so.

To begin with, it should be reminded that most of the scholars, who have either taught mathematics or written mathematics textbooks, have at the same time been philosophers, theologians, astronomers, natural scientists and, very often, clergymen. As will be discussed later, it is in those other capacities of the scholars, rather than their capacity as mathematicians, that laid the cause of the conflicts created at times.

In the first centuries after the fall of Constantinople, the Church replaced the lost empire in the consciousness of Greeks and became the political instrument of the cohesion of the enslaved Greek nation[1]. The Patriarch, thanks to privileges granted by Mehmed the Conqueror, assumed the role of the political and religious leader of the Greek-Orthodox people[2]. During the following centuries, in particular, Eastern Orthodoxy was almost identified with the national idea. Even after the rise of the Phanariotes, the Church continued to be a very powerful institution and the supreme political form of the Greek nation, despite the permanent threat of Islam, on the one hand, and Catholic propaganda, on the other[3]. As it could therefore be expected, it supervised the sociopolitical affairs of the Greeks as well as every aspect of education.

The decision of the 1593 Synod, under Patriarch Jeremiah the Great, on the establishment of schools in the provinces, initially boosted education among the Greeks (although, eventually, the decision was rendered inactive). The Patriarchal Academy and the Athonias School were the only schools to be founded on the initiative of the Ecumenical Patriarchate. However, all schools established in the broader Greek intellectual territory — thanks to donations by the Greek merchants of the Diaspora or the princes of Wallachia and Moldavia or the personal initiative of members of the clergy — obtained the patriarchal sigil and, in essence, were supervised directly by the Patriarchate.

Undoubtedly, the Church played a positive part in local educational issues, particularly up to the 17th century. Since then, however, its contribution was significantly limited by the effort to undermine the exact sciences, an effort which was essentially a manifestation of the Church's bewilderment and fear of the newly imported Western European spirit. The representatives of the Church correctly understood that the study of exact sciences constituted a fundamental part of a new intellectual orientation and that the spread of the exact

1. N. Svoronos, " Preface ", *A Review of Greek History*, Athens, 1992 (in Greek).

2. *Idem*, 44.

3. *Idem*, 49.

sciences did not aim solely to the enrichment of the Greek people's knowledge, but also, to the creation of a new life outlook and to rational thinking. Clearly, Church circles did not desire this.

Up to the 18th century, the works of Aristotle were, according to the Church, the only pure source of scientific knowledge. From the early 17th century until the early 18th century, Theophilos Korydaleas transplanted the new perspective on Aristotle's philosophy in Greece. It is Aristotle's works proper, and not anymore just the commentaries of paraphrasts and editors, that were considered as the only starting point for the instruction of philosophy courses in schools. The curriculum was no longer confined to logic and rhetoric. There was a notable production of interpretive commentaries on Aristotle's works (*Metaphysica*, *De Caelo* and *Meteorologica*), on various subjects, ranging from physics to metaphysics. Nikolaos Mavrokordatos was the first to doubt the " absolutization " of Aristotelism by the Eastern Orthodox Church circles, on issues relevant to research about nature and the study of human values. He compared Aristotle to contemporary exact scientists in his work *Philotheos's Sidelines* (1709)[4]. Thus, in the early 18th century, the Patriarchate started receiving different messages from Phanari, which had already evolved into a second authority center. However, it kept the gates to the West shut, in order to maintain control over the curricula of higher educational institutions in the broader Greek space, not forgetting the traumatic experience of its acquaintance with Calvinism by K. Loukaris[5], Patriarch in the early 17th century.

There was always a justifiable fear of Catholic propaganda and there were efforts to confront it. Let us not forget the Flagginiano Tutorial School in Venice (a school established by Thomas Flagginis) and the Kotouneio *Ellinomouseio* (school) in Padova, which were established and functioned as a counterweight to the College of St. Athanasios in Rome, founded in the 16th-17th century by the Roman Catholic Church, and the Catholic School in Chios, with its 250 students. Natural sciences and mathematics were returning to Greece via the West. This return path could, in itself, be expected to yield the suspicious response of the Patriarchate and " conservative circles ". With the passing of time, there was a growing awkwardness by the Patriarchate circles and the clergy concerning the Enlightenment movement and, consequently, the exact sciences. Eventually, the West was identified with the idea of atheism, which was to become the most common accusation against Latin oriented (or not, occasionally) scholars. It should, however, be noted, that this disbelieving behavior toward the import of modernist attitudes and perceptions was not

4. N. Mavrokordatos, *Philotheos's Sidelines*, see N. Psimmenos, *Greek Philosophy*, vol. II, 43 (in Greek), [" ... at the same time, I admire and cannot stop praising the modern (scientists) ... even the wise Aristotle ... would want to be a student of such men... "].
5. N. Psimmenos, *Greek Philosophy*, vol. II, *op. cit.*, 15.

always adopted by the official Church, but very often by an overly conservative section of clergy or laymen, who believed that scientific progress conflicted with orthodox faith and feared that science would upset the closed system of unquestioned faith which they had created.

About the end of the 17th century rose the first instances of a conflict between conservative circles and Church representatives on the one side, and scholars trying to introduce the modern western spirit, mainly in the field of education, on the other side. One such instance was the case of Georgios Sougdouris, who became headmaster of the Giouma School in Ioannina in 1683. This Italy-educated scholar introduced exact sciences in the school and systematically taught Philosophy using new methods, "causing a storm of protest". The bishop of Ioannina at that time, Klimis, asked for the scholar's excommunication because Sougdouris was not thinking according to divine rules[6]. In the same period, Papavassileiou attempted to hint, for the first time, at mathematics, following European models. "But his teaching was condemned for being 'atheist' and Papavassileiou stopped teaching mathematics, fearing excommunication by the Church"[7]. Thus, in both cases mentioned above, the scholars were accused of atheism.

The first clear clash between the ruling Church and sciences was the case of Methodios Anthrakitis, although accusations against the scholar concerned the teaching of Molinos's theory, or rather, the questioning of Aristotelian philosophy. In the scholar's own words "I am therefore condemned by the Synod, not for being a bad Christian ... but for philosophizing in a way different from the Aristotelians"[8]. According to A. Aggelou (a scholar of nowadays, who died in 2000) this was perhaps the first restraining intervention of the Church in the content of scholarly teaching at that time[9]. A point worth mentioning in the case of Anthrakitis is the fact that his condemnation was co-signed by Chrysanthos Notaras, one of the most luminous minds in the Church[10].

In 1753, Iosipos Moisiodax requested the support of the Church of Smyrna toward the completion of his studies in Padova. Another manifestation of the fear of imported atheism and its possible results can be seen in the fact that the headmaster of the Evangelical School of Smyrna, Ierotheos Dendrinos, discouraged the scholar's plans, holding that all who studied in the West were

---

6. K. Sathas, *Neohellenic Literature, Biographies (1453-1821)*, Athens, 1868, 394 (in Greek).

7. Ph. Michalopoulos, *Ioannina and Neogreek Revival (1648-1829)*, Athens, 1930, 35 (in Greek).

8. A. Aggelou, *Epiphany*, Athens, 1988, 26 (in Greek).

9. *Idem*, 23.

10. M. Gedeon, *The Intellectual Movement of the Nation in the 18th and 19th Centuries*, Athens, 1976, 100 (in Greek), (" ... condemned him in August, with the agreement of Chrysanthos of Jerusalem ").

atheists[11]. Later, Moisiodax was accused of being western-minded, because of his preference for European education. His characterization as a " heterodox " was clearly just a step before accusing of atheism.

Several of this period's scholars were subject to the ecclesiastical circles' wrath, at the end of the 18th century and the beginning of the 19th.

The " modernistic " ideas of the headmaster of the Patriarchal Academy from 1809 to 1810, Stephanos Dougas, who taught mathematics, physics and philosophy, using the new methods, were not in accordance with conservative circles. Dougas, " unable to fight countless machinations, left the school after just one year (1810) "[12]. In 1815, Dougas submitted his book with the title *Investigation into Nature* for approval by the Patriarchate, a process to which several scholars were subjected. He obtained a negative response and was accused of being a heretic by Dorotheos Voulismas, as, according to the latter, in the book, the spirit had a material existence (" lacking spirit, he speaks of spirit... "). Dougas was forced to make a confession of faith. The scholar was again accused of atheism by the Patriarchate, during his service as headmaster of the Jasy Academy, and was forced to resign[13].

Accusations of atheism against Veniamin Lesvios, instructor of scientific courses in the Kydonies Academy, were presented to the Patriarchate by its circles in 1805 with no immediate effect, but approximately ten years later, the expected result was achieved : Lesvios had no alternative, but to resign[14]. As stated by Veniamin in his apology to Kallinikos v of Nikaia : " I am not an instructor of theology, but of mathematics and natural sciences, therefore there cannot be blasphemy in mathematics as a science, even if the instructor is the worst of human beings[15] ". Thus, in his response, Lesvios stated a truth, which the Patriarchate, hemmed in by fear and prejudice, could not conceive.

Other, though not very different, negative features attributed to the sciences in general, and mathematics in particular, by their opponents, was the percep-

11. I. Moisiodax, *Apology*, Vienna, 1780 (reprinted Athens, 1992), 153 (in Greek).

12. K. Koumas, *Histories of the Acts of Mankind*, vol. XII, Vienna, 1832 (reprinted Athens, 1966), 593 (in Greek).

13. Ar. Camariano-Cioran, *Les Academies Princieres de Bucharest de Jassy et leurs Professeurs*, Thessaloniki, 1974, 653-655.

14. G.P. Henderson, *The Revival of the Greek Thought 1620-1830*, Athens, 1977, 169-1999 (in Greek).

15. Sophoklis Oikonomos, *Literary Works of Constantine Oikonomos*, vol. I, Athens, 1871, 435-437. See M. Gedeon, *The Intellectual Movement of the Nation in the 18th and 19th Centuries*, *op. cit.*, 120 (in Greek). Gedeon mentions " Veniamin, incessantly subjected to machinations and accused of atheism to the Church, came to Constantinople, in order to defend himself in person against hateful slander, and proved beyond doubt the purity of his faith ".

tion that sciences were useless and dispensable[16]. Evidently, these features were indirectly linked to the accusation of secularization of religious belief, in the sense that the main duties of Christians are prayer and repentance and not occupation with matters that distance them from religion. Essentially, therefore, this characterization also fell within the " sciences are an agent of atheism " framework.

Diabolical features were, moreover, often attributed to the science of mathematics[17]. Breaking the fast and disrespect for religious ritual in general, were considered results of the atheism, which was brought about by mathematics[18].

In the course of the broader ideological opposition of representatives of tradition, expressed mainly by the patriarchate environment to scientific knowledge and mathematics (an opposition which resulted in efforts to preserve control over school curricula), scholars of proved religious belief were also persecuted. One of them was Nikiphoros Theotokis, a great scholar and clergyman of the 18th century, who had to leave the Jasy School, where he served as schoolmaster (1765), " by night, like a fugitive "[19], following problems created by conservative circles. A fundamental element in Theotokis's lawfulness was his cautious attitude toward the theory of the heliocentric system. In his *Physics*, published in 1766, he presented the heliocentric system. Yet, the expressions he used, e.g. " they hypothesize that the earth moves " and " weak hypothesis ", which placed Copernicus's system in the sphere of hypothesis, as well as his scientific repute, provided support for several opponents of the heliocentric system, which was fought against by the Church. The latter had, for centuries, accepted the geocentric one, which was harmonious with theological affairs.

16. N. Psimmenos, *Greek Philosophy (1453-1821)*, vol. II, 426 (in Greek). In an excerpt from a letter of deacon Makarios of Patmos to Neophytos of Arta, referring to the case of Anthrakitis, we read : " ... it seems that Methodios teaches his students triangles and squares and the rest of the vanity of Mathematics... " .

17. M. Gedeon, *The Intellectual Movement of the Nation in the 18th and 19th Centuries, op. cit.,* 104 (" From the mid-18th century and very sadly for the sciences, like in the Byzantine years, when people also happened to believe that their initiates into civilization were spirits which had escaped from hell and sons of the devil ").

18. A. Parios, headmaster of the Chios School, wrote in his *Response* (1802) that " in those years, anyone who set foot in Europe was, without further examination, an atheist... mathematics was the source of atheism, the first result of which was breaking the fast ", and referred to Varlaam Kalavros as insane. *Response to the Phrenetic Zeal of the Philosophers who Come from Europe*, 1802, 50, 68-70 (in Greek).

In the encyclical issued by Patriarch Grigorios V in 1819, one of the reasons for his bitter attack against mathematics is " disrespect for the fasting ecclesiastical rules and unresponsiveness to the ritual of our immaculate belief and worship ". *Ilissos*, 30th June 1870, Athens, 97 (in Greek).

19. K.Th. Dimaras, *Neohellenic Enlightenment*, Athens, 1983, 77 (in Greek).

However, in order not to present a disproportionately negative picture of the Church, it should be noted that the persecutions of scholars that took place at times varied in intensity. There were cases of complete extermination, both moral and physical, such as the case of Th. Kairis (whose extermination, however, was realized in the mid-19th century by the authorities of the newly established Greek state). Most cases, however, involved simple recommendations or requests for a document of confession of faith.

In March 1819, Patriarch Grigorios V issued an encyclical, in the form of a synodical letter, " on the current state of the common *Ellinomouseia* (schools) of our Nation ". Among other issues addressed in the encyclical, Grigorios attributed the abandonment of the " instruction of supreme Theology ", the disregard for " literary courses " and the rest of what he considered the evils of Greek education, to the " devotion of students and teachers solely to Mathematics and the Sciences ", which was propagandized by some " corrupt excuses for men ", who " grew like pests... among the educated people of the Nation ". Scholars who taught mathematics were referred to as " offenders of divine and holy rules ", " cunning enchanters " etc. The exact sciences were not considered beneficial, but rather, harmful to youth, while it was stated that those who suggest that youth study mathematics " maliciously prevent their (youth's) progress through true education "[20]. It was also noted that " Algebra, cubes, cubes cubed, triangles and triangle squares, logarithms, symbolic numbers, atoms ... " will bring about atheism. The 1819 encyclical was the starting point for the decline and disintegration of the pre-Revolutionary educational institutions, which had been inspired by Enlightenment ideology[21]. It was complemented by the Patriarchal Synod on 27th March 1821 " on the removal of philosophical courses ". Decisions taken included the immediate termination of the operation of " modernistic " schools, mainly those of Kydonies and Smyrna, the prohibition of the instruction of philosophical courses (the term included mathematics) and the expulsion of Constantinos Koumas and Veniamin Lesvios from Smyrna because " they did not serve the interests of philosophy "[22]. Theophilos Kairis and Neophytos Vamvas had the same fate[23].

It should be also noted that the language in which some of the offended scholars responded to criticism was certainly not less harsh. Stephanos Oikonomou referred to those he considered responsible for the polemics against him

20. *Ilissos, op. cit.,* 99, 100.

21. F. Helios, *Blind, Lord, thy People*, 623-624 (in Greek).

22. K. Oikonomos, letter to St. Oikonomos, 17th September 1821, see K. Lapas, " Patriarchal Synod on the removal of philosophical courses on March 1821 ", *Mnimon*, vol. XI (1987), 125 (in Greek).

23. K. Lapas, " Patriarchal Synod on the removal of philosophical courses on March 1821 ", *op. cit.,* 142.

in the Literary Gymnasium of Smyrna as "envious and self-interested"[24] "conspirators" and "werewolves". Even Korais used analogous language, naming those who viewed European wisdom as atheism "male crones"[25].

An interpretation of the Patriarchate's attitude, particularly in the last decades prior to the Greek revolution, is attempted below, through a brief reference to conditions in this period.

It is known that, until 1791, the Patriarchate operated under the influence of two opposing forces. On the one side, there was the Ottoman Empire, to which it had to demonstrate submission[26], while on the other side, there was the Russian Empire, with which it sought to retain warm relationships, as the Russian Empire often supported Greeks and adhered to the same religious dogma. However, Catherine II promoted the development of philosophical ideas, willing to prove her liberal spirit on the one hand, but often came into conflict with Ottoman interests, on the other hand. In these conditions, the Patriarchate had to keep a delicate balance, in order not to displease any of the two sides.

From 1791 onwards, however, the situation changed. Catherine renounced liberal ideas and the war with Turkey was over. Close relationships with Voltaire discontinued and she was followed in this by Eugenios Voulgaris, her adviser and guide in the policy of friendliness with the Orthodox peoples. Possibly, the change in the Patriarchate's attitude was partly due to these facts. No longer fearing the Empresses's disapproval, the Patriarchate could accuse and attack modernistic spirit with relative ease, as its attitude and the climate in the centralized Russian court converged. The nastiest works against philosophers and scientists were published in those years.

In the last decade prior to the Greek Revolution, the vision of spiritual rebirth started being realized, particularly in the field of education, with the operation of "modernistic" schools and the gradual decline of more traditional ones. The introduction of new methods and ideas evidently led to questions over the soundness of the educational system, which had heavily depended on scholarly tradition and had always operated under the wings of the official Church. This rebirth was indissolubly connected to the exact sciences, which were introduced and taught in order not only to produce new

---

24. Rescued literary works, see Chr.S. Solomonidis, *Education in Smyrna*, Athens, 1961, 47 (in Greek).

25. Ad. Korais, … *Conversation between two Greeks* (in Greek). He writes " they have directly moved from the foolish age of children to the more foolish age of old men, without ever becoming perfect men, because they never obtained a man's mind. Those I name male crones, and believe them to be much more foolish than the female ones ".

26. See P.G. Metallinos, *Turkish Domination*, Athens, 1993, 157 (in Greek) " the Turkish authority worried for affairs both domestic and external, and evoked encyclicals by the Constantinople Patriarch, which excommunicated Greeks publishing liberal works abroad " (I. Filimon).

practical knowledge, but also to contribute to the revival of theoretical thinking and the creation of a new attitude to life[27]. Certainly, this was a result of the Enlightenment movement, whose agents in the Greek intellectual space were growing in number. The liberal ideas of the French Revolution had been spreading beyond the borders of France, and the conservative Patriarchate circles, always accountable to the Sublime Port, feared the consequences of the penetration of these ideas among the Greeks. The prospect of an impeding revolution, following the French model, seemed particularly worrying, as conservative circles believed that such an attempt would be disastrous for the Greek nation. The French Revolution, which demonstrated the dangers of the new ideas for the ruling class, marked the end of the somewhat " tolerant " spirit which had, until then, been manifested by the ruling Church. The clash was severe. It is known that most persecutions were realized in the last two decades prior to the Greek Revolution. Until the French Revolution, as long as western ideas seemed restricted to a level of the introduction of new scientific knowledge, there was no serious clash, in the scientific sphere, between the Orthodox Church and the representatives of Enlightenment. However, as soon as the danger was felt at different levels, those of ideas, the clash moved into a sector where the Church would surely lose : the field of exact sciences, *i.e.* the specific scientific knowledge itself, and not the field of the confrontation of this knowledge[28].

The last act of the drama, *i.e.* the Synod decision, in March 1821, on the removal of teachers and the termination of the operation of modernistic schools, aimed to extinguish all breeding grounds which could be used as a pretext for the Patriarchate to be accused, by the Sublime Port, of fostering the revolutionary movement. It was essentially a symbolic act, by which the Patriarchate would prove its conformity to the Sublime Port and its complete opposition to the modernistic ideas, which were identified with the spirit of rebellion[29]. It is known that prior to the Revolution, one of the accusations stated against the enlightened scholars was that their innovatory perceptions supported a climate which was favorable for the development of revolutionary ideas.

---

27. N. Svoronos, " Preface ", *A Review of Greek History, op. cit.,* 56.

28. E. Nicolaïdis, " Orthodoxy ", " Religious Humanism " and " Enlightenment ", *Neusis,* issue no. 1, Fall, 1994, 119 (in Greek). A typical example of this attitude was the theory of heliocentrism, which was introduced with a delay of approximately a century. Other examples are the work of Sergios Makraios, *Trophy of the Greek Armor Against the Supporters of Copernicus in Three Dialogues,* Vienna, 1794, and the re-edition of the old manuscript of Voulgaris's *Universal System,* Vienna, 1805 (in Greek).

29. K. Lapas, " Patriarchal Synod on the removal of philosophical courses on March 1821 ", *op. cit.,* 139, 141.

The above presentation is mainly founded at the level of personal cases, simple reference to data and presentation of facts. The only assertion, which will be attempted, is that the Church's distrust of mathematics was not primary, but was rather determined by the diversity of activities pursued by those who occupied with mathematics, who had and expressed new perceptions on society, state and religion, essentially questioning the leading role of the Patriarchate. There is no evidence of dispute on any purely mathematical issue containing heresy. It is worth quoting again Veniamin Lesvios' words : " ... it is impossible for blasphemy to exist in mathematics, even if the instructor is the worst of human beings "[30]. Therefore, mathematics was included in the broader questioning and distrust of the exact sciences which, being a product imported from the West, were considered as a fundamental agent of atheism and a factor destabilizing the dominant order in ecclesiastical, as well as national, issues. During the pre-Revolutionary century, supporters of both sides, believing in the holiness of their cause, reached extremes in passion unmet in the history of the Greek communities of the Ottoman Empire[31], but essentially, the clash was due to the misinterpretation of each other's intentions.

30. See footnote 15.
31. K.Th. Dimaras, *Neohellenic Enlightenment, op. cit.,* 307.

" RENEGADES " AND MISSIONARIES AS MINORITIES IN THE TRANSFER OF KNOWLEDGE

Sonja BRENTJES

Missionaries are famous as transmitters of knowledge in early modern times with regard to China, Southern America, and — to some extent — India. Less well known are their activities in Africa and Western Asia. The earliest efforts to print books for missionary purposes, however, focussed upon Middle Eastern languages, mainly Arabic and Syriac, and included books related to the sciences such as Ibn Sina's medical *Canon*, a version of Euclid's *Elements* today known as Pseudo-Tusi, and a fragment of al-Idrisi's geography, then known as *Geographia Nubiensis*. These well-known facts suggest to assume that missionaries did play a certain, even if only minor, role in the transfer of knowledge between Western Europe and the Middle East during the late 16th and the 17th centuries.

In the first part of my article I will show that while the missionaries indeed played only a minor role in the transfer of knowledge their activities covered a wide variety of topics. The second part of my article will present some evidence for the participation in the transfer of knowledge of a different group of migrators between Western Europe and the Middle East — the so-called " renegades ". With regard to them, early modern sources often claim that they were of key importance for all that was modern in the Ottoman Empire and its Northern African dependancies — whether technology or science or art. On the basis of my sources, I will argue that contrary to these claims, the so-called " renegades " also exercised only a minor impact upon the transfer of knowledge — excluding, however, technology — between Western Europe and the Middle East[1].

---

1. Some of my vocabulary remains by necessity ambiguous such as " transfer of knowledge ", " Western Europe ", or " Middle East ". I use the first because the networks through which knowledge was transmitted in early modern times were only partially connected with universities or *medreses* and because the transmitted knowledge was partially produced outside these two major teaching institutions. That is why I wished to apply a notion which is not strictly disciplinary bound. I employ the two other designations as anachronistic shorthands for the Catholic and Protestant countries of Europe on one hand and for the Ottoman Empire, the Safavid Empire, and the kingdom of Morocco, on the other hand. The focus of my talk, however, is — with regard to the Middle East — upon the Ottoman Empire.

MISSIONARIES

Missionaries came in the late 16th (Jesuits) and the early 17th centuries
(Capuchins, Carmelites) to the Ottoman Empire[2]. They erected schools for
boys of the Christian communities where Arabic, Greek, Italian, and Catholi-
cism were taught[3]. They also went to teach these matters in schools run by the
various Christian communities themselves[4]. Language and religion were the
main fields were the Propaganda fide and the Missions Etrangères contributed
as institutions to the transfer of knowledge between Western Europe and the
Middle East. They organized the printing of grammars, dictionaries, apologies,
the new testament, and prayer books. The printed grammars and dictionaries
were derived from Oriental antecedents which had been brought by merchants,
Oriental Christians, or Western European scholars to Rome or Paris. Before
they could be printed they had to be checked for errors and reproduced as
ready-made copies. In this process, as a rule, Oriental Christians were actively

2. In 1631, the Jesuit Jérome Queyrot gave the following survey of the orders working in the
Ottoman Empire : *Outre les Pères Observantins qui servent de chapelains aux Français et aux
Vénitiens qui ont leur chapelle séparément, et outre nous qui ne sommes que deux, sans frère ni
serviteur, il y a des Capucins et des Carmes réformés. Les Capucins sont cinq prêtres et un lai.
Les Carmes sont trois prêtres et un frère. On nous a donné, depuis peu, espérance de dresser dere-
chef une école, mais sans nous séparer, comme nous fîmes au commencement. Les Pères Capucins
se mêlent encore d'enseigner, non seulement ici où ils ont six écoliers, mais encore à
Constantinople ; et en peu de temps ils se sont fort multipliés en Levant. Car, outre la maison de
Constantinople et d'ici, ils se sont encore logés à Chio, à Smyrne, en l'île de Naxie, à Saide, à
Parut* (sic) *qui sont en Syrie, et en Alexandrie d'Egypte. Il sont allés aussi en Babylone, qu'on
nomme maintenant Bagdad ; mais à cause des guerres qui sont entre les Persans qui ont pris cette
ville-là aux Turcs, et le Grand Seigneur de Constantinople, les bons Pères ont été contraints de
l'abandonner.* quoted after : Rabbath, 1907, 381-382.
3. These schools were erected in the early 17th century and are referred to regularly in the various reports by the superiors of the Jesuits and Capuchins in Syria. In 1629, Jesuit Jérome Queyrot
wrote from Aleppo to the French ambassador in Istanbul : *Par celle que j'écrivis à V. E., ce
carême passé, vous aurez appris l'heureux commencement et progrès d'une école dressée, depuis
mon arivée en cette ville, dans la maison du Métropolite des Grecs, chez lequel je demeure, ...
Cette école va croissant, de jour en jour, si bien que l'on y compte maintenant jusques à trente
enfants Grecs, qui apprennent en grec, en arabe et en italien. Nous espérons qu'avec le temps
ceux des autres nations se serviront de nous, comme les Grecs, en l'nstruction de leurs enfants, si
Notre-Seigneur nous fait la grâce d'avoir ici quelque maison, comme il est du tout nécessaire.*
quoted after : *idem*, 380. 24 years later, in 1653, the Jesuits had opened schools in other Arabic
towns too. The Jesuit Jean Amieu, for example, wrote from Beyrut to Rome : *Tripoli autem scho-
las Græcorum obibat, pueros ad elementa pietatis erudiens, nec intermittens ipse per se domi,
Sidone ante et Alepi, ætatulam teneram adhuc formare ad bonos mores, instituendo ad primordia
litterarum, superior et professus quatuor votorum, imo missionum omnium curam gerens.* quoted
after : *idem*, 428. Three years later, the Capuchin Alexandre de Saint-Sylvestre reported to Paris :
*En cette résidence d'Alep, il y a ordinairement quatre religieux ; Le Frère de notre Dame du Mont
Carmel, Arménien de nation, tient école en notre maison, apprenant à lire et à ecrire en langue
arabesque et en langue italienne à vingt jeunes enfants, tant catholiques que des autres sectes susdites, et les instruisant en la doctrine chrétienne, et aux mystères de notre sainte foi.* quoted after :
Rabbath, 1907, 434.
4. Alexandre de Saint-Sylvestre stated : *Là, le Père Anselme de l'Annonciation, Français de
nation, va le matin et le soir aux écoles publiques des schismatiques enseigner le catéchisme et la
doctrine chrétienne ; ce qu'il fait encore en d'autres maisons particulières, tant des catholiques
que des sectes susdites, où plusieurs personnes s'assemblent avec la même affection ; ...* quoted
after : Rabbath, 1907, 435.

involved. In the Ottoman Empire, missionaries worked on manuscripts written in the local languages hoping to publish them after their return to Europe[5]. The Carmelites Ange de Saint Joseph and Ignace de Jésus composed multi-lingual encyclopedic dictionaries such as *Gazophylacium linguae Persarum* started by the former in Basra in 1666 and published in Amsterdam in 1684 or the Latin-Arabic and Latin-Persian dictionaries by the latter which remained unpublished[6]. Within this frame, some missionaries also begun inquiries in languages and peoples either badly known or altogether unknown in Europe. Ignace de Jésus, Mathieu de Saint Joseph, and Ange de Saint Joseph worked on the language, culture, and geographical distribution of the Mandaeans who were regarded in Europe as the so-called St John's Christians. Ignace published in Rome in 1652 a book on this matter. The book was accompanied by a map of the Euphrates-Tigris delta which presumably was based on an Arabic predecessor extant today in a private collection[7]. The works of Mathieu and Ange remained unpublished[8].

The only other form of institutionally backed participation of missionaries in the transfer of knowledge was the transport of books and instruments to the Ottoman Empire in order to be capable to refute views of the Oriental Christians and in order to attract members from those as well as from Jewish communities and induce them to convert. In 1639, the Jesuit Aimée Chezaud after having described his progress in learning Arabic, informed Mutio Vitelleschi, the General of the Society of Jesus, that he had asked Athanasius Kircher to send several books needed by the missionaries such as the *Compendium totius Linguae Arabicae Thesauri*, the Arabic Grammar by Alvarez, and a Turkish-Latin dictionary[9]. The Jesuit Adrien Parvilliers took instruments to the Ottoman Empire which he used to show off to Jewish and Greek youngsters :

J'avais commencé d'attirer chez nous quelque jeunesse d'esprit, par le moyen de quelques verres et gentilles de mathématique, à dessein de les apprivoiser à entendre parler de nos mystères. ... Je les ai gagnés et faits amis, non-seulement par les caresses et civilités avec lesquels je les reçois, mais encore par le moyen des instruments de mathématique comme globes, sphères, cartes, verres triangulaires, dont ils sont curieux[10].

5. Nicolas Poiresson wrote, for instance, about Jérome Queyrot : ... *le Père travaille à plusieurs livres arabes et grecs vulgaires pour leur faire voir le jour en leur temps.* Rabbath, 1907, 68.

6. See, for instance, Bastiaensen, 1985.

7. See *The History of Cartography*, 1992, 222-223. I was informed about the sale of the map by the manager of the Bernard Quaritch Ltd., the former owner.

8. See Bastiaensen, 11-12.

9. Rabbath, 1907, 93.

10. *Idem*, 61-62.

Aside from such institutionally backed activities, a number of missionaries undertook on an individual level observations of nature and the heavens while in the Ottoman Empire. They measured geographical latitudes or quoted them from books, described a variety of events such as earthquakes, extraordinary falls of snow, unusual excitments of the Mediterranean sea, and other phenomena which they considered as portents, a belief they shared with their Muslim contemporaries whose calendars they studied. Such information was submitted to the superiors at home as, for instance, did Nicolas Poirresson in his *Relation des Missions de la Compagnie de Jésus en Syrie en l'Année 1652.* sent to France :

> *Pour ce qui est des turcs en tous ces pays, tant qu'ils seront les maîtres, il y a peu ou rien à espérer. Je ne sais pas ce que Dieu veut faire de changement en ces pays, mais cette année, que j'ai été ici, est remarquable en ses prodiges, comme les commencements de 1653, bien marqués, disent les turcs dans leur calendrier, et y attendant de grandes révolutions. Ces prodiges ont été vues sur terre, sur mer, en l'air et au ciel, à la grande Mosquée de Hiérusalem, bâtie sur le temple de Salomon*[11].

These reports also served to describe the geography of the regions where the missionaries worked and traveled as well as their history and contemporary political, social, or economic situation. Such descriptions enumerated distances between major cities and ports, listed rivers and mountains, named plants and animals, and talked about public buildings and commercial goods[12]. While the missionaries embroidered the description of their work in the Ottoman Empire by observations of nature and society, in the Empire itself they took recourse to the sciences and scientific instruments in order to gain the attention of Muslims, Jews, and Christians alike. The already mentioned Adrien Parvilliers was highly regarded by the literate elite and the common folk in Sayda because of his excellent command of Arabic, his ability to read the books which even the *shaykhs* of the mosques understood only by half, and because of his knowledge in astrology and mathematics which gained him high credit among the " Turks " and the Greeks. He predicted an eclipse and talked to the locals about other heavenly events and their causes[13].

While these activities were not censured by the superiors and thus can be regarded as at least tolerated, other activities of individual missionaries were looked upon with less than approval. These activities, however, appear to have constituted the bulk of what the missionaries contributed to the transfer of knowledge. They concerned the exercise of medicine on the one hand and the

11. Rabbath, 1907, 71.
12. *Idem,* 65-66.
13. *Idem,* 59.

cooperation with members of the Republic of Letters in Western Europe on the other hand.

The missionaries in their zeal to save as many souls as possible often went to see sick people, in particular women and children. They consoled them with prayers and baptized those who were dying. While such activities were part of their duty, they tended to include the administration of remedies to improve the state of the bodies too. The exercise of medicine was, as a rule, forbidden to the orders. A special dispence by the Holy See was required in order to practice it legally. Such a dispence received among the orders working in the Middle East only the Theatins whose activities were mainly in the Caucase mountains. All the other orders were repeatedly criticized by their superiors in Rome or Paris for engaging in such illicit actions[14].

Members of the Republic of Letters such as the French Nicolas Fabri de Peiresc or the Dutch Jacob Golius approached Capuchins and Carmelites in the Ottoman Empire time and again to acquire Arabic, Persian, Coptic, Greek, Syriac, or Ethiopian manuscripts needed for their various studies. The missionaries, in order to fullfil these requests, cooperated with Muslim scholars, bibliophiles, book-traders, and scribes as well as with Syriac, Greek, and Coptic priests and scribes. Peiresc also talked several missionaries into astronomical, geographical, and historical observations and excursions. Such activities were often regarded by the superiors as detrimental to the missionaries' religious duties and devotions.

Missionaries could not simply take off to observe ecclipses, collect plants or stones, copy inscriptions, or hunt for rare manuscripts. Before they could leave their hospices they had to acquire written permissions either by the superiors in the Middle East or, if the planned travel was of greater length and distance, from Rome. Father Celeste of St Lidvine, brother of Jacob Golius, after having been introduced by Peiresc into the adventures of observing nature and the heavens took such a fascination in them that he begged Peiresc to get him a special license from Rome for further undertakings[15].

The evolution of a more stable cooperation between scholars in Western Europe and missionaries in the Ottoman Empire was hindered by the often relatively short period of time which the missionaries spent in a particular town in the Ottoman Empire. After the missionaries had learned Arabic in Aleppo

---

14. See, for instance, Rabbath, 1907, 400, footnote 1, who wrote that *les certaines compositions de médecines aisées* mentioned in a report by Father Amieu *ont toujours été en usage, parmi les missionaires de tous les Ordres. Ces œuvres de miséricorde corporelle leur ont ouvert bien des portes et bien des cœurs, et ont considérablement augmenté le nombre des petits baptisés qui se sont envolés au ciel, d'entre les bras de leurs parents infidèles ou négligents.*

15. See Ms Paris, BNF, Dupuy 688, ff. 19a-22b.

they were sent away to other towns, within the Empire and outside of its borders. While the communication with major towns in the Ottoman Empire in form of letters, parcels, or travels was relatively stable and took, as a rule, only a few months, the communication with missionaries in Iran or India took much more time and money. Although individual scholars like Peiresc received a few letters from India too, those were mostly written by merchants, not by missionaries. The contacts of the missionaries working in Iran or India were limited to the reports for their superiors, to letters written to their co-fathers, and to letters to the French ambassador. They concerned above all matters of religion, politics, and private life.

" RENEGADES "

" Renegades " are not well studied because there are not many sources available which make them visible. They appear rarely if ever in Muslim literature, while they can be found more often in European sources. If they are mentioned, either their previous life as Christians or their later life as Muslims remains, as a rule, opaque. An example is Katib Çelebi's information about his partner, Mehmed Ikhlas, a former French priest with whom he worked on rendering Gerard Mercator's *Atlas Minor* into Ottoman Turkish and who is said to taught him how to dress maps in a European style[16]. While Katib Çelebi mentioned Mehmet Ikhlas repeatedly in his works and thus provided us with all the information about their joint venture which we possess today, he said nothing about his former life, not even naming the town where he had been born, let alone the order to which he had once belonged. The " renegades " mentioned in Western European travel accounts are mostly pirates from the North African coast or dignitaries on various levels of the Ottoman administration. Whether these " renegades " could have contributed substantially to the transfer of knowledge is difficult to evaluate. In order to be capable of doing so, they must have been born in Western Europe and must have passed at least a college education, if not a university training. One case of such a " renegade ", called Osman d'Arcos, can be found among the correspondents of Nicolas Fabri de Peiresc. Before he converted to Islam in about 1634, he was Thomas d'Arcos, the former secretary of cardinal Joyeuse. Since 1630, he was a prisoner in Tunis waiting for being ransomed. After his ransom had arrived, Thomas d'Arcos decided to convert and stay there for the rest of his life.

Peiresc who never accepted d'Arcos' conversion engaged him in an intensive exchange of letters, manuscripts, animals, plants, and sweets. He inquired about Punic and Roman historical sites, bones of giants, ecclipses, tides, currents, and winds, informed the " renegade " about interviews he himself had

---

16. See, for instance, Katib Chelebi, 1957, 144.

carried out with captains, merchants, and scholars investigating these topics from the Atlantic coast of Spain to the Bosphorus, and tried his best to make d'Arcos participate in his vast program of observation by pointing out to him that major, most important insights could be gained by his cooperation[17]. D'Arcos, on his side, was mostly interested in winning Peiresc's approval for his own writings about Africa, history, creation, politics, and law. He sent them to Aix-en-Provence, discussed with Peiresc their relative merits, and received books such as the Latin translation of the so-called " Geographia Nubiensis " from Peiresc in order to incorporate some of its information into his own work on Africa[18].

Osman d'Arcos was not the only " renegade " who contributed to satisfy Peiresc's vast curiosity in nature and humankind. D'Arcos himself delivered information about apes collected by a " renegade " from Ferrara in his capacity as court officer responsible for the trade caravane to Niger and Mali. French merchants in Cairo such as Jean Magy and Gabriel Fernoulx informed Peiresc about news they had received from various Muslim sources, among them Muhammad Pasha, the then governor of Suakin, a convert from Ragusa who had received his training at the serail in Istanbul[19]. The information concerned the outbreak of an Ethiopian vulcano, earthquakes in Cairo, extraordinary weather appearances and their devastating results in Mecca, and other phenomena[20]. It was valuable to Peiresc since he was studying vulcanic activities, their connection with subterranian caves, the influence of winds upon such activities, and related questions.

I am convinced that other material confirming the involvement of converts to Islam in the mutual transfer of knowledge during the 16th and 17th centuries can be found as is proven by a manuscript extant in the Staatsbibliothek Berlin composed presumably by a " renegade " who knew Greek, Slavonic, Italian, Persian, and Arabic[21]. Their transitory nature as converts, however, will always pose an obstacle to fully uncover their respective contribution.

BIBLIOGRAPHY

Bastiaensen, Michel (ed.), *Ange de Saint-Joseph, dans le siècle Joseph Labrosse. Souvenirs de la Perse safavide et autres lieux de l'Orient (1664-*

---

17. Tamizey de Larroque, vol. 7, 1898, 128, 142. Ms Carpentras, Bibliothèque Inguimbertine, 1821, f. 127a.
18. Tamizey de Larroque, vol. 7, 1898, 109-110, 115, 125 ; Les Correspondants de Peiresc, vol. 2, 1972, 208, 215.
19. Ms Carpentras, Bibliothèque Inguimbertine, 1864, ff. 263a-b, 264a.
20. *Idem*, ff 263a-264a.
21. See Ms Berlin, Staatsbibliothek, Preussischer Kulturbesitz, Or. Oct. 33.

1678), Faculté de Philosophie et Lettres XCIII, Editions de l'Université de Bruxelles, 1985.

Les Correspondants de Peiresc. Tome II. *Lettres Inédites publiées et annotées par Philippe Tamizey de Larroque*, Genève, Slatkine Reprints, 1972.

The History of Cartography, vol. 2, book 1, in J.B. Harley (ed.), Chicago, London, The University of Chicago Press, 1992.

Katib Chelebi, *The Balance of Truth*, Translated by G.L. Lewis, London, George Allen and Unwin ; New York, The Macmillan Company, 1957.

Rabbath, Pere Antoine, *Documents inedits pour servir à l'histoire du Christianisme en Orient (XVIe-XIXe siècle)*, Paris, Leipzig, London, 1907.

Tamizey de Larroque, Philippe, *Lettres de Peiresc*. Tome Septième. *Lettres de Peiresc à Divers. 1602-1637*, Paris, Imprimerie Nationale, 1897.

MANUSCRIPTS

Ms Carpentras, Bibliothèque Inguimbertine 1821.

Ms Carpentras, Bibliothèque Inguimbertine, 1864.

Ms Paris, BNF, Dupuy 688.

Ms Berlin, Staatsbibliothek, Preussischer Kulturbesitz, Or. Oct. 33.

OTTOMAN ENGINEER MEHMED SAID EFENDI AND HIS WORKS ON A GEODESICAL INSTRUMENT (*MÜSELLESIYE*)[1]

Mustafa KAÇAR and Atilla BIR

It is in the military field that one witnesses the first applications of the movements of change and modernization emerging in the Ottoman State in the early eighteenth century. Ulufeli Humbaracılar Ocağı (The Corps of Salaried Bombardiers) established in 1735, formed the beginning of this movement. Mehmed Said Efendi, with whose work we shall deal in this study, was an Ottoman "engineering teacher" who taught geometry at the Corps. As a multi-faceted scholar he becomes the subject matter of our article, as he had worked on the solution (by means of an engineering instrument he claimed to invent) to the significant problem pertaining to distances that could not be measured.

For many years it has been generally accepted that Ottoman science repeated itself through commentaries and explanations, that it closed itself to European influences and that it was unaware of the modern sciences developing in Europe. However, for the last two decades new information has been found and new theories have been developed as a result of the studies carried out by the Department of the History of Science at the University of Istanbul, by the Research Center for Islamic History, Art and Culture (IRCICA) in Istanbul. In this regard, especially *History of Ottoman Astronomy Literature, (OALT) History of Ottoman Mathematical Literature (OMLT) and History of Ottoman Geography Literature (OCLT)* edited by Ekmeleddin Ihsanoğlu and published by IRCICA within the project of Ottoman Scientific Literature are a treasure for those who do research in these fields.

1. This work was supported by The Research Fund of the Istanbul University : project n° : B-985/3105201.

The present study deals with one of the studies by Mehmed Said Efendi, son of an Ottoman mufti. We hope that Said Efendi's studies as well as his work titled *Rub'-ı Müceyyebü'l- Zülkavseyn* will contribute to the enlightenment of a new aspect of the Ottoman scientific world.

His Life and Family

Although there is not much information about his life, we find through his own records, some information about his family, especially his children. Said Efendi, son of the Mufti of Beyşehir[2], Elhacc Mahmud b. el-Hacc Hasan b. el-Hacc Ahmed, had, just like many Ottoman authors and intellectuals did, the habit of writing down on the covering page of the books he authored some information on his family, birth and some events he considered important. He wrote down especially his children's dates of birth, in both Islamic and Christian calendars, including the exact hours and minutes of their birth. Besides he did not neglect to record their birthplaces and horoscopes[3].

We learn from Said Efendi's records that he married twice and had six children. He married his first wife Şerife Hanım, daughter of Esseyid Ali Ağa on 15 Safer 1145 (7 August 1732). He stated that this fell on 26 July, but did not give the year. His first child Şerife Rukiye was born on 26 Zilkade 1145 (10 May 1733). He informs us that this daughter was born in April. Yet she passed away in Cemaziyelahir 1147 (Ekim 1734) when she was only two years old. He had from Şerife Hanım, a son called Seyyid Abdullah in 1147 (1734), but he too passed away in the same year. Thus Said Efendi lost both his children in the same year.

Said Efendi's second wife was Fethullah Hanym, daughter of Mestçizade Abdullah Efendi, son of Anatolian Kazasker Osman Efendi. From this wife he had four children who lived long. The first child was a daughter called Ayşe and born on 17 Safer 1153 (14 Mayıs 1740). She was born in a house near the tomb of Seyyid Buhari in the district of Sultan Mehmed. The second child from Fethullah Hanım was Mahmud Mes'ud who was born on 3 Receb 1154 (14 September 1741). Said Efendi gave this date as 3 September 1741[4]. Fatma

2. Mehmed Said Efendi is usually stated as the son of the Mufti of Yenişehir rather than of Beyşehir. These results from confusion, as the name of the two towns were written quite similarly. When his works studied carefully it becomes clear that he was the son of Hacı Mahmud Efendi, the Mufti of Beyşehir. E. Ihsanoğlu, *OMLT*, vol. I, 180 ; A. Adıvar, also stated that Said Efendi was the son of the Mufti of Beyşehir, see *Osmanlı Türklerinde Ilim*, Istanbul, 1984, 183-184.

3. Mehmed Said Efendi, *Resâîl-i Saidiyye*, (Evahir-i Şaban 1152/1739) Library of Topkapı Palace Museum (TSMK), Hazine, n° 1753, 78 folios.

4. There is a 11-day difference between the Christian date Said Efendi provided and the date according to modern calculations ; see Faik Reşit Unat, *Hicrî Tarihleri Milâdî Tarihe Çevirme Kılavuzu*, 6th ed., Ankara, 1988.

who was born in Beykoz on 6 Zilhicce 1156 (21 Ocak 1744) on a Tuesday was the third child. And the last child was Mehmed Esad who was born on 29 Receb 1158 (27 August 1745) on a Saturday.

Said Efendi definitely put his name and the date the work was completed at the end of each work of him. This seems to be something differentiating him from other authors. Thus we learn about the dates of the completion of his works as well as his duties. In only one place do we come across his personal seal inspired by the stars. The octagonal seal, which is provided below, reads as follows : *Hemvâre necm baht ola Mehmed Said* (Mehmed Said, may whose star always rise)[5].

The Establishment of the Corps of Salaried Bombardiers

Among the duties of Mehmed Said Efendi, the one about which we have the most information is his teaching position at the New Corps of Bombardiers. The Corps of Salaried Bombardiers had a significant place at the beginning of the modernization movement in the Ottoman Empire.

When Mahmud I (1730-1754) was enthroned on 1 October 1730, amidst the chaos in the political and military scene in western Europe and endless Iranian expeditions in the east, he tried to improve the Ottoman military situation and sought alliances with European powers. Again, he initiated the first efforts towards a reformation of the Ottoman army in line with the European model.

It is seen that Ottoman reference is towards Europe, in establishing a new system whose bases are outside the classical Islamic world, that is, in Christian Europe. The Corps of Salaried Bombardiers became the first example of such an approach. It was established in 1735 under the supervision of the French General Claude Alexandre de Bonneval (1675-1747) who took refuge in the Ottoman Empire in 1729. Comte de Bonneval, who was known to Ottomans as Humbaracı Ahmed Paşa, became well known for his successes in the fields of military technique and art of war. He had a significant place in the Ottoman history due to his successes in the use of modern warfare techniques as well as the application of Western innovations to the military field[6].

5. Süleymaniye Library, Esad Efendi n° 3704, folio 1[b].
6. M. Kaçar, " Osmanlı Devleti'nde Askeri Sahada Yenileşme Döneminin Başlangıcı ", *Osmanlı Bilimi Araştırmaları* (I), in F. Günergun (ed.), Istanbul, 1995, 209-225.

By the imperial decree of 25 January 1835, the establishment of a new Corps of Bombardiers was approved and the relevant regulations were issued. Thus the classes of bombardiers were reorganized and the corps was composed of 301 salaried bombardiers. Bonneval Ahmed, who became a pasha with the rank of *mîr-i mîrân,* was appointed as the Chief Bombardier of this corps. First of all, the officers were selected, their salaries were determined and a hierarchy was established within the corps. Accordingly, the first commander-in-chief (*alay başı*) and 300 bombardiers were organized in three divisions *(odas)* and for each division a group of 25 officers were assigned. These officers displayed some differences in accordance with their divisions, yet their number remained the same.

The reform movement starting with the aim of establishing a new type of Ottoman military power, which would gain supremacy against Europe, emerged in the form of the provision, both in theory and application, of a new military education by a European general like Bonneval Ahmed Paşa. As mentioned above, the reform was carried out within and in accordance with the existing classical military organization, but differed in content from other similar military corps. The main distinguishing characteristic of the Corps of Bombardiers was that its members were educated in mathematics and geometry and that they were made to be experts in firearms.

Although the Corps of Bombardiers did not live long and lost its function a few years after its establishment, the notion of a new military formation embodied in this first attempt displaying a certain European influence was important in the sense that it served as a model for engineering schools to be established in the last quarter of the 18th century.

Education at the Corps of Bombardiers

One does not find any information regarding the establishment, in the corps of bombardiers, of an *engineering school* or any institution providing engineering education. Nevertheless, the fact that there were among the officers such staff as engineering teacher (*hoca-yı mühendis*), technical drawing teacher (*muallim-i resim*) and teacher of the division (*hoca-yı oda*) shows that several courses on theoretical mathematics, geometry and practical engineering were given in the corps[7].

7. Prime Minister Ottoman Archives (BOA) *Maliyeden Müdevver* (M.MD), n° 5941, 48.

We know that apart from Said Efendi[8], who was an engineering teacher at the barracks of bombardiers, there were among the teaching staff Ali Ahmed Hoca (from Kasımpaşa)[9] with a daily wage of 40 *akçes*, Süleyman b. Hasan (from Istanbul)[10] with 40 *akçes* and Osman b. Abdullah[11]. We see that the technical drawing teacher was Ybrahim of Istanbul who left in 1152/1739 his post with 60 *akçes* per day[12]. The other instructor (*hoca*) was probably Said Efendi's friend Abdullah el-Muzafferî el-Bosnavî.

Selim who was *ser-çavuş* of the corps with a daily wage of 240 *akçes* was at the same time the teacher of firearms mathematically and technically. Selim, a convert, probably of French origin, was also " the Chief Architect of War ". Engineer Selim who had received an education in the field of war engineering since his youth was appointed under Bonneval Ahmed Paşa as Chief Sergeant of the first division of the Barrack of Salaried Bombardiers. He was an expert in the fields of construction of new castles, preparation of artillery and bombardier bastions as well as technical drawing pertaining to the military architecture[13].

The Work of Said Efendi

Said Efendi's bibliography is provided in detail in *OALT*[14] and *OMLT*[15]. We have his works in two collections. He wrote the first one during his teaching career at the Corps of Salaried Bombardiers dated 1149/1736 (Süleymaniye Library, Esad Ef. n° 3704). The second one is his *Resâil-i Saidiyye* dated 1152/1739 (Library of Topkapı Palace Museum, Hazine n° 1753). In both works there are many treatises of him, some of which are written in verse.

One of the most interesting characteristic seen in Said Efendi's treatises is that he employs two different styles. While sometimes he uses a plain language and simple explanations, at other times he provides more advanced geometrical solutions. Another striking characteristic of his studies is the numerousness and the diversity of the topics he dealt with. To call him a geometry teacher who

---

8. His name is mentioned in 1152/1739 at the second line of the absentee list of the Corps of the Bombardiers ; BOA, *M.MD.* n° 5941, 48.

9. His name is mentioned in 1739 at the sixth line of the absentee list of the Corps of the Bombardiers, BOA, *M.MD.* n° 5941, 48.

10. His name is mentioned in 1739 among the present at the Corps ; BOA, *M.MD.* n° 5941, 42.

11. A professor of the first division at the Corps in 1147/1735, BOA, *M.MD.* n° 5941, 12.

12. His name is mentioned in 1739 among the present at the Corps, BOA, *M.MD.* n° 5941, 42.

13. See the margin. BOA, *M.MD.* n° 5941, 110.

14. Vol. II, 458-461.

15. Vol. I, 212-214.

knows astronomy well would not be wrong. His works which we shall cite below show clearly the multi-facetedness of the Ottoman scholars of the classical period.

In the introductory parts of his works, Said Efendi states that in the past Muslims were stronger that Christian powers, but that for some time " the people of the north ", that is Europeans made advances in firearms and thus gained supremacy over the Muslims. Therefore, in line with the common idea of the time, he emphasized that the art of war was necessary for victory and that knowledge of engineering was essential to this art.

The first *Mecmua* (Süleymaniye Library, Esad Efendi N° 3704) consists of Said Efendi's writings on a variety of issues. The first treatise is about constructing vertical sundials deviating from the east-west direction. This work is studied in the following article of the present book : *Ottoman Engineer Mehmed Said Efendi and his Treatise on Vertical Sundials.* Another work is about the determination of distances which cannot be calculated without using an instrument of measurement. This treatise was probably written in order to be used in the classroom when he taught at the Corps of Bombardiers[16]. The work which is in Arabic and has two different diagrams deals with one of the problems of the classical Islamic geodesy[17].

The second *Mecmua* which is at Topkapı Palace Museum Library titled *Resâil-i Saidiyye* related to engineering consist of following treatises

1. *Rub'-ı Müceyyebü'l-Zülkavseyn* (Sinus quadrant with double arcs) (vr. 18b-37a) which will be dealt with in detail.

2. The treatise on the usage of Sinus quadrant with double arcs prepared on order of the Grand Vizier of the time, El-Hacc Ahmed Paşa. (vr. 37b-43a)

3. *Pergel-i Nisbenin Imali ve Bazı Burhanı* (The construction and proof of proportional pair of compasses) (vr. 68a-71b).

Said Efendi wrote about matters other than engineering and engineering instruments as well. Among them was a book titled *Tilavet*, which he wrote in order to help his daughter Ayşe to pronounce correctly when she recited the Holy Qu'ran. In this work, which was said to be the first in Turkish, he shows, with the aid of diagrams, that tongue, larynx and the structure of the teeth are influential while reciting the Qu'ran[18].

16. At the end of the treatise he signed as " Mehmed Said Mühendisi-i Ocağ-ı Humbaracıyan " (Mehmed Said, Engineer of the Corps of the Bombardiers), *Fasl-ı Mesaha bilâ Alat* (in Arabic) Sül. K. Esad Ef. N° 3704, 241.

17. Donald R. Hill, *Islamic Science and Engineering*, Edinburgh, 1993, 199-200.

18. M. Said Efendi, *Hediyetü'l-Talibîn fî Ta'lîmi'l-Kur'ânu'l-Mübîn*, TSMK, Hazine, 1753, fols. 58b-67b.

Said Efendi's Personality

Being a member of the *Ilmiye* (men of learning) class and having a degree of scholarship (*müderrislik*) at *kırklı medrese* (the teacher receives 40 *akçes* per day), Said Efendi was a multi-faceted scholar who wrote in " three languages " (Arabic, Turkish and Persian) in such fields as mathematics, geometry, astronomy, sundials, surveying and terrestrial globe as well as recitation of the Qu'ran. Based on the work we are studying here, one can say that he was meticulous in his works. While the problems he chooses as examples are well-chosen and didactic, no error is found in their solutions.

Said Efendi who was a family man tried to learn French and Bosnian as well[19]. He often exchanged ideas with converts. He also had some inventions. The last date mentioned about Said Efendi is 18 Safer 1171 (19 September 1760). Bursalı Mehmed Tahir writes in his *Osmanlı Müellifleri* that " Mehmed Said Efendi of Istanbul who was the time-keeper of the Sultan Mustafa (Laleli) Mosque enlarged and translated in 1181/1767 Müneccim Musa b. Hasan Nevbaht's work dealing with issues related to the movements of stars "[20]. On the other hand, as Çınarizade (Halifezade) Ismail Efendi was appointed as the time-keeper of the same mosque in 1181/1767, one may conclude that he was Said Efendi's successor and Said Efendi passed away in 1767[21].

The Treatise on the Usage of the Instrument Sinus Quadrant with Double Arcs (*Rub'-ı Müceyyebü'l-Zülkavseyn*) Called *Müsellesiye*

Müsellesiye was a geodesy instrument designed for calculating the long distances which could not be measured directly. One part of the instrument fixes the angles while another part fixes the trigonometrical functions of those angles in relation to side lengths[22]. It was an instrument designed for the calculation of the length of the two sides of a triangle whose length of one side

19. We see that from time to time Said Efendi systematically writes the Turkish equivalents of the French and Bosniak words, thus we can conclude that he tried to learn these languages. Süleymaniye Library, Esad Efendi n° 3704, 379.

20. Bursalı Mehmet Tahir, *Osmanlı Müellifleri*, vol. III, 272.

21. *OALT*, vol. II, 530-536. The late Cevat Izgi mentions briefly Said Efendi in his *Osmanlı Medreselerinde Ilim*, vol. I, Iz Yay., Istanbul, 1997, 314.

22. On sinus quadrant see, Atilla Bir, " Zamanı Belirlemeye Yarayan Aletler ", in ed. Kazım Çeçen, *Osmanlı Imparatorluğu'nun Doruğu*, Istanbul, Istanbul Büyükşehir Belediyesi (ISKI) yay., 1999, 265-269.

and the degree of two angles were known. Said Efendi wrote his *Sinus Quadrant with Double Arcs* on the construction and usage of this instrument[23].

The work seems to have be completed in two stages. He could complete only in 1737 the " circle " he designed for calculations in 1735. He showed the instrument first to Mehmed Pirizade Efendi, the Chief-Judge (*Kazasker*) of Rumelia and was praised by the latter. Through him and with the help of *Silahdar Ağa* he had the opportunity to present his invention to Sultan Mahmud I who in return awarded him with mid-level *medrese* degree and who appointed him to Kadı-zâde Ahmed Medrese.

Said Efendi's " the new instrument " or *müsellesiye* was formed from two connected mechanisms both of which were devices for conventional measuring and calculation and had been used in the Muslim world for various purposes. One was a sinus quadrant, (*rub'-ı müceyyeb*) while the other was a telescope (*durbîn*)[24]. Said Efendi's new invention had the following advantage : it brought the two devices together, enabled more sensitive calculations as it used two telescopes and could be used easily by the bombardiers. Although the origin and the qualities of the telescope were not mentioned in his work, it is known that at that time Ottoman engineers used telescope imported from Europe for border measurements[25].

a) Parts of the Instrument

Muhit : The big circle forming the *Müsellesiye*.

Izade : Two movable arms.

Fuls : Two metal discs.

Nutak : Sockets on the arms.

Sar : Covers on the sockets.

---

23. Author manuscript of this treatise is at TSMK, n° : Hazine, 1753, v. 18b-36b. It was copied in 1211/1796 as *Humbara Risalesi*. A copy whose copyist is not known is registered at Büyükşehir Belediyesi, Taksim Atatürk Library, Muallim Cevdet collection, n° K. 145. Except for some writing errors, this fine copy with its coloured drawings was very helpful in studying the work. According to A. Adıvar, another copy is registered at Berlin Library, Pertsch catalogue, n° 166. Adıvar states that as he could not find the treatise in the libraries of Istanbul, it was not possible to understand clearly what was the invention (mentioned in the work) about. A. Adıvar, *Osmanlı Türklerinde Ilim*, 183.

24. D.R. Hill, *Islamic Science and Engineering*, op. cit., 195-197.

25. We know that during the drawing of the new Ottoman-Austrian borders after the Treaty of Belgrade in 1840, Eğinli Numan Efendi watched by a binocular the works of the Austrians and made a similar instrument of measurement with which he revealed errors committed by them. This incident was mentioned first by Erich Prokosch, " Osmanische Geodäsie um die Mitte des 18. Jahrhunderts ", *60 Jahre Höhere Technische Bundes-Lehr- und Versuchsanstalt Mödling 1919-1979*, Mödling. We would like to thank him here for kindly sending us this article. Prof. Prokosch has another work on Numan Efendi and the drawing of the Austrian border : *Molla und Diplomat*, Graz, 1972. Cevat Izgi dwelt on this issue as well : *Osmanlı Medreselerinde Ilim*, 314-318.

Sehpa : Tripod.

Mihver : Nail.

Dürbün : Two telescopes.

Zemin terazisi : Balance.

Hayt : Colored strings (6 pieces in different colors).

The instrument consists of a circle called *muhit* and is divided into four equal parts. Each of this quarter-circle is divided into grades (from left to right and vice-versa) according to the *abjad* system. On the nail there are two height set-squares attached to each other at the center. One side of the set-squares coincide with the diameter of the circle. The two sides of the set-squares divide the circle into two equal parts. There is a binocular in each of the sockets called *nutak*, located parallel to each of the set-squares. These binoculars are directed at the target. The looking part of the field glass is called *reis* (head) and the ruler parts, *zeneb* (tail). Under both set-squares are two discs (*fuls*), one with a red rope, the other with a green one. All these are fixed from two sides of the circle by means of an object called *fers*.

Sinus quadrant with double arcs is the quarter-circle with sine lines, inside of an other bigger quarter-circle. The radius on its right is called *mütemmim* (cosine radius) and the one on its left *ceyb-i tamam* (sine radius). Both are divided into 85 degrees and they stats from the center.

The quarter-circle is called altitude arc and is graded starting from cosine radius. Its polar point (*kutup noktası*) is a hole at the 60th grade of the sine radius. Here is a yellow string called polar string. The sections divided between the pole and the center are sine radius and the line descending from the pole to the 45th degree of the altitude arc is called hexagon *sittinî*. It is divided into 60 degrees and its beginning is at the polar point. The section between the polar point and the altitude arc is called *zaid* (plus) and is divided into 15 degrees. The arc drawn on the pole between the end of the hexagon and the center is called *kavs-ı dahil* (inner arc) and is divided into 90 degrees starting from the center. All the degrees mentioned above are on the circumference and subdivided in 6 minutes.

The lines descending from the cosine radius to the altitude arc are called *ceyb-i menkus*, while those from the sine radius *ceyb-i makus* ; and the lines from the hexagon reaching the inner arc are called *cüyub-ı mebsuta* (tangent). The beginning and number of these lines depend on the corresponding degrees.

b) Usage of the *Müsellesiye*

The *nâzır* side (looking point) of the distance to be measured is called *menşe* first point (M), while its *mer'i* side (the point seen or the target) is named *muhit*. The tripod is put at first point *menşe*. Müsellesiye is located exactly in the middle of the tripod. With the aid of the hydrostatic balances it is assured that *müsellesiye* is properly located.

With the aid of the hydrostatic balance an object which can be seen through *vasat* (second point) which is located at a desired distance towards the appropriate direction from the plummet direction of the tripod. The amount of distance is recorded and it is accepted as the known side of the triangle. Then one field glass of the *müsellesiye* is directed towards the *muhit*, while the other towards the object at the *vasat* point, and the movable arms are moved until the *muhit* and *vasat* point become visible. Thus an angle is formed between the target and the known side. The degrees covered on the *muhit* by the *münassıfs* which are between " head " and " tail " parts of the movable arms become the angle of the *menşe* point, and here is put a second object which can be seen at the plummet direction of the tripod.

The tripod is carried to the *vasat* point and, as done before, *müsellesiye* is located properly exactly in the middle of the tripod. The same operation of measurement is repeated and this time the angle of the *vasat* point is determined. Thus we have a triangle with the measurements of whose two angles and one side are known. Then the lengths of the other two sides and the distance of the target point are found.

Said Efendi ends this section with a note of " importance " : " It should be known that when these two angles are added and the sum is subtracted from 180°, then the result is the angle of the *muhit*. Relevant diagrams and drawings are provided at the end of the treatise which continues with the section of purpose (*maksad*) where the calculation of the lengths of the two unknown sides are explained.

c) Exemplary problems

Said Efendi takes six different examples and suggests one or more solutions for each of them. In the examples, first of all, he gives, the theory, then the proof and finally, the solution. As mentioned earlier, the solutions are errorless. However, because in some trigonometrically functions he takes only four digits after the comma, some very minor differences occur among the results.

The examples he takes are carefully chosen and enable the user to do the easiest and most correct measurement, for often the user can determine the *menşe* angle by himself. Said Efendi gives, at the beginning, operations for measurements whose *menşe*, *vasat* and *muhit* angles are 90 degrees. Secondly, operations are given for obtuse angled measurements, and finally for acute-angled ones. And at the appendix part of the work, for determining the distance between two targets in front, he provides three different solutions for three different angles.

Some examples and their solutions

All measurements in a triangle depend on the principle of knowing the angles and one side of that triangle (Figure 1).

Angles of the Triangle :

$M = Menşe'$ (first point),

$V = Vasat$ (second point),

$H = Muhit$ (target).

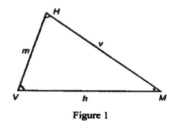

Figure 1

Sides of the Triangle : $VM = h,$ $HM = v,$ $VH = m.$

Measured angles ; M and V, then $H = 180° - (M + H)$,
the supposedly known side of the triangle : h,
the unknown sides : m and v.

1 - When the *Menşe* (or *vasat*) is a right angle ($M = 90°$) (Figure 2)

According to the sine theorem ;
$m/\sin M = v/\sin V = h/\sin H$,
or $\sin M = \sin 90° = 1$ and $\sin H = \sin (90° - V) = \cos V$,
then,
$m = v/\sin V = h/\cos V, m = h/\cos V$,
$v = m \cdot \sin V = h \cdot tg\ V, v = h \cdot tg\ V$,
can be deduced.

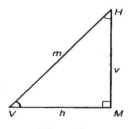

Figure 2

a) First Method (Figure 3)

The string is put on angle V that is measured before and the mark M_1 put on the opposite of $OC_1 = h$ then $OS_1 = v$, $OS_2 = OM_1 = m$.

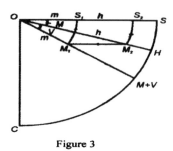

Proof :

OC_1M_1 is a right triangle
$\cos V = OC_1 / OM_1 = h/m$,
    $m = h/\cos V$,
$\text{tg } V = C_1M_1 / OC_1 = v/h$, $v = h \cdot \text{tg } V$.

Figure 3

Example :

If $h = 20$, $V = 70°$, $M = 90°$, $H = 90° – 70° = 20°$, then :
$m = h/\cos V = 20/\cos 70° = 58,5$,
$v = h \cdot \text{tg } V = 20 \cdot \text{tg } 70° = 54.9$.

b) Second Method (Figure 4)

The string is at the angle $H = 90° – V$.
The mark M_1 is brought opposite to
$OS_1 = h$, then $OC_1 = v$, $OS_2 = OM_1 = m$.

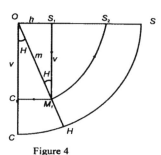

Proof :

OS_1M_1 is a right triangle
$\text{tg } H = OS_1/S_1M_1 = h/v$ and
$\text{tg } H = 1/\text{tg } V$ therefore,
$v = h/\text{tg } H = h \text{ tg } V$, $\sin H = OS_1/OM_1 = h/m$
and $\sin H = \cos V$,
$m = h/\sin H = h/\cos V$

Figure 4

Example :

If $h = 20$, $V = 70°$, $M = 90°$, $H = 90° – 70° = 20°$, then :
$m = h/\sin H = 20/\sin 20° = 58,5$, $v = h/\text{tg } H = 20/\text{tg } 20° = 54,9$.

2 - When the *Muhit* is a right angle ($H = 90°$) (Figure 5)

VHM is a right triangle ; $m = h.\sin M$,
$v = h.\cos M$.
The string is at the angle M (Figure 1).
The mark M_1 is set on $OS_2 = h$,
then $OS_1 = m$, $OC_1 = v$.

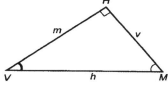

Figure 5

Proof :

From the right triangle OM_1C_1
$\sin M = m/h$ and $\cos M = v/h$
can be written.

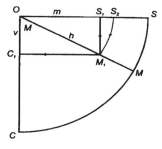

Figure 6

Example :

If $h = 58\ 1/3$, $V = 20°$, $M = 70°$, then :
$m = h.\sin M = 58\ 1/3.\sin 70° = 54,81$
$v = h.\cos M = 58\ 1/3.\cos 70° = 19,95$

3 - When the *Menşe* is an obtuse angle ($M > 90°$) (Figure 7)

When the sine theorem is applied to the
HVM triangle, the following equations
can be written :
$h/\sin H = v/\sin V = m/\sin M = m/\sin (180°$
$- M) = m/\sin (H + V)$.
Specially $m = h.\sin (H + V)/\sin H$ and v
$= m.\sin V/\sin (H + V)$.

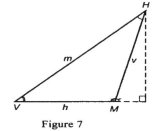

Figure 7

To find the side m (Figure 8) :

The mark M_1 on the first string is set on
$OS_1 = h$. The string is brought to the angle
$(H + V) = 180 - M$. The mark M_2 on the
second string is set so that
$M_1M_2 /\!/ OS$. Then $OM_2 = OS_2 = m$.

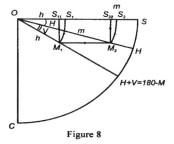

Figure 8

To find the side v (Figure 9) :

The mark M_2 on the second string is set on to $OS_2 = m$. The string is brought to the angle V. The mark M_1 on the first string is set so that $M_1M_2 // OS$. Then $OM_1 = OS_1 = v$.

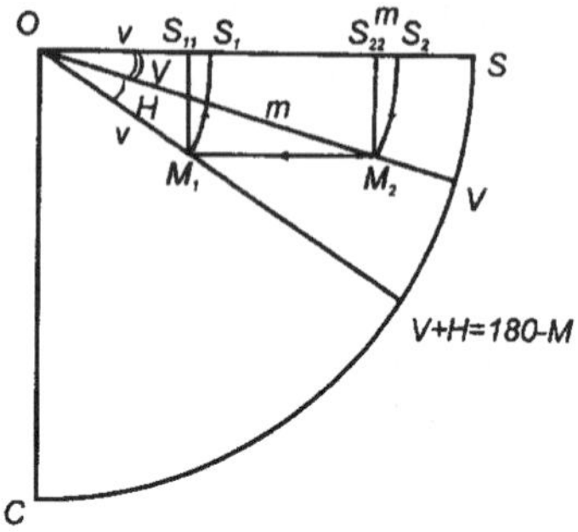

Proof :

In (Figure 8) $OM_1\ S_{11}$ is a right triangle ;
$\sin (H + V) = M_1S_{11} /\ OM_1 = M_1S_{11}/h$.
OM_2S_{22} is a right triangle ;
$\sin H = M_2S_{22}/OM_2 = M_1S_{11}/m$.
When M_1S_{11} is eliminated in both equations ; $m = h\ .\ \sin (H + V)/\sin H$.
In Figure 9, OM_1S_{11} is a right triangle ;
$\sin V = M_2S_{22}/OM_2 = M_2S_{22}/m$.
OM_1S_{11} is a right triangle ; $\sin (H + V) = M_1S_{11} /\ OM_1 = M_2S_{22}/v$.
When M_2S_{22} is eliminated in both equations, $v = m\ .\ \sin V/\sin (H + V)$.

Figure 9

Example :

If $h = 20$, $M = 150°$, $V = 17°$, $H = 13°$, $V + H = 30°$, then :
$m = h\ .\ \sin (H + V)/\sin H = 20\ .\ \sin 30°/\sin 13° = 44{,}45$,
$v = m\ .\ \sin V/\sin (H + V) = 44{,}45\ .\ \sin 17°/\sin 30° = 29{,}99$.

4 - When the Menşe is an acute angle ($M < 90°$) (Figure 10)

From the sine theorem one can write ;
$v/\sin V = m/\sin M = h/\sin H = h/\sin [(180° - (V + M)] = h/\sin (V + M)$
or $m = h\ .\ \sin M/\sin (V + M)$, $v = h\ .\ \sin V/\sin (V + M)$.

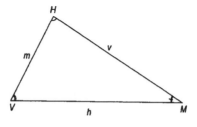

Figure 10

Like in item 3, the calculation of side m is started with the angle M and the known side h (Fig. 11) ; side v is calculated by changing the sequence of the angles (Fig.12).

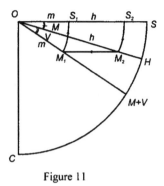

Figure 11

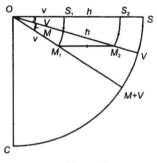

Figure 12

Example :

If $h = 44,45$, $M = 13°$, $V = 17°$ $V + M = 13° + 17° = 30°$, then :
$m = h \cdot \sin M/\sin (V + M) = 44,45 \cdot \sin 13°/\sin 30° = 20$,
$v = h \cdot \sin V/\sin (V + M) = 44,45 \cdot \sin 17°/\sin 30° = 25,99$.

5 - When the *Muhit* is an acute angle ($H < 90°$) (Figure 13)

When $H < 90°$ or $V + M > 90°$, the construction given in item 3 and 4 are not applicable, since the intersection occurred outside the quadrant. In this case, one use the hexadecimal second quadrant and the string on its center.

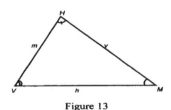

Figure 13

a) First method (Figure 14) :

Given that side h and angles V and M are known ; the angles $(V + M) – 90°$ and $90° – M$ are calculated. The first angle is set on the big quadrant and the second on the small one. The marks on both strings are placed on the point M_1 of intersection. $O'M_1$ string is turned around O' so that the mark M_1 coincides with M_1'. Side h is added to M_1' point to obtain $M_1'M_1'' = h$ and M_2' is set on this point and the string is moved back to its previous position. Then following the line $M_2'M_2// OS$ the second mark is set on M_2. The distance measured by moving the string OM to the side OS, gives the distance $M_1'''M_2''' = m$. From

M_2' and M_2 parallel lines are drawn to side OC, so that on OS line S_1' and S_1 points are obtained. The distance between S_1' and S_1 is equal to the distance v ; $S_1'S_1 = M_2'M_2 = v$.

Proof :

The triangle $M_1M_2'M_2$ is
similar to triangle which we
attempt to find.
$M_1M_2' = M_1'M_1'' = h$ and
$M_2'M_2 // O'S'$ implies : angle
$M_1O'S'$ = angle $M_1M_2'M_2 = M$.
M_1 is an external angle of
OM_1O' triangle ;
M_1 = angle $O'OM_1$ + angle
$OO'M_1 = V$
and $M_2'M_2 = S_1'S_1 = v$.

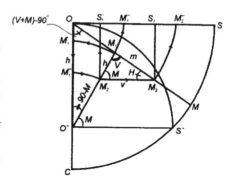

Figure 14

Example :

If $h = 20$, $M = 60°$, $V = 80°$, $H = 40°$, $(V + M) - 90° = (80° + 60°) - 90° = 50°$,
$90° - M = 90° - 60° = 30°$, then :
$m = h. \sin M/\sin H = 20$ $\sin 60°/\sin 40° = 26{,}95$,
$v = h. \sin V/\sin H = 20$ $\sin 80°/\sin 40° = 30{,}64$.

b) Second Method (Figure
15) :

In this case, in the big quadrant angle H is marked starting from side OS and in the
small quadrant angle V is
marked starting from side
OS'.
As in case a), from triangle
$M_1M_2'M_2$ the angles and the
sides are obtained as ; angle
$M_1 = M$, angle $M_2' = V$, angle
$M_2 = H$,
$M_1M_2 = v$, $M_2'M_1 = h$
and $M_2'M_2 = m$.

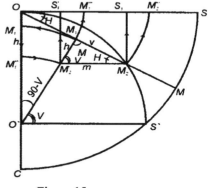

Figure 15

Conclusion

It is known that after its invention in Europe in the 16th century, the tele-scope was used for measuring long distances as well in addition to the pur-poses of observation. The telescope, which was used by Ottoman sailors in the 17th century and imported widely in the 18th century, was favored by the Otto-man palace. One sees that the telescope was used for the first time as a scien-tific instrument in 1737 instead of the conventional *rub'u tahtası* (quadrant) used for measuring distances and heights. In the present article we have tried to study Mehmed Said Efendi, one of the Ottoman engineers from the early era of modernization, and the *müsellesiye* which he claimed to invent. This instru-ment composed of two different devices, an Ottoman scientific instrument (*rub'u tahtası*) and a product of European technology (telescope), embodies at the same time a synthesis of two different cultures. Said Efendi does not pro-vide information on the telescope he used, nor how he obtained them. Like-wise, he does not feel a need for explaining *Rub'-ı Müceyyeb* which was widely known and used in the Ottoman world.

By bringing together two different devices which are independent from each other, yet used for the same purpose, Said Efendi invented a new instrument for land surveying which did not exist neither in Europe nor in the Ottoman world. In fact what made this instrument different from the similar instruments was that the measurement is made by constructing on the instrument itself a triangle similar to the hypothetical triangle whose one side is the distance to be measured. A second characteristic of the instrument is that the user can find the required distance easily without the aid of any other measuring instrument or calculation.

A Drawing of the *Müsellesiye*

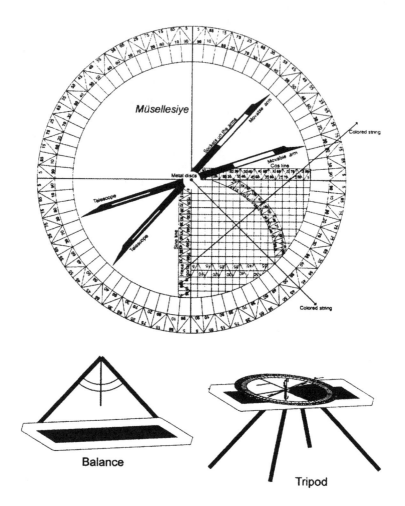

Original Drawing of the *Müsellesiye*

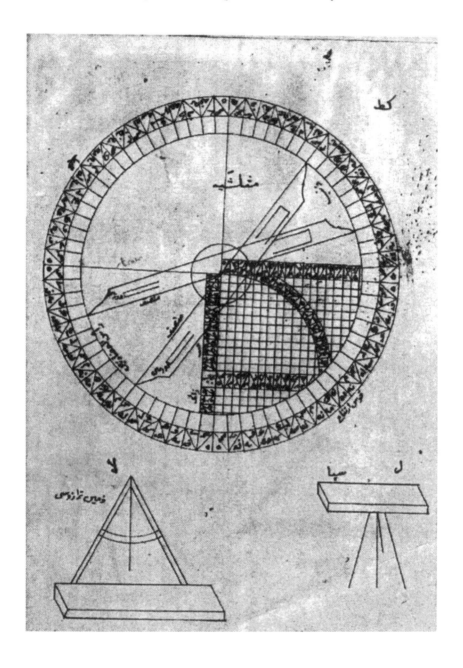

OTTOMAN ENGINEER MEHMED SAID EFENDI AND HIS TREATISE ON VERTICAL SUNDIAL

Atilla BIR and Mustafa KAÇAR

We have dealt with the life and works of Said Efendi, an 18th century Ottoman engineer, in the previous article of the present book : " Ottoman Engineer Mehmed Said Efendi and His Works on a Geodesical Instrument (*Müsellesiye*) ". And in the present article, we shall study his work *Duvara Inhiraf-ı Saat-ı Nehar istihrac etmenin tariki* (the construction of vertical sundials deviating from the east-west direction) and its explanatory appendix *Medarat Vaz' etmenin Tariki* which he authored in 1737.

Said Efendi wrote *Duvara Inhiraf-ı Saat-ı Nehar istihrac etmenin tariki* which was about the determination of the deviation degree of vertical sundials from the east-west direction in 17 May 1737. He had heard about this problem from his colleague Abdullah Efendi el-Muzafferi el-Bosnavi[1], an instructor at the Bombardier Corps in 1735[2].

---

1. There is not much information on Abdullah Efendi el-Muzafferi el-Bosnavî. His name tells that he was from Bosnia and he was probably one of the bombardiers brought from Bosnia during the establishment of the Corps of Salaried Bombardiers. Abdullah el-Muzafferî, possibly a convert, came to Istanbul in 1147/1735 and left in 1149/1737. His name is mentioned among the leaders of fifty (*ellibaşı*) with 90-*akçes* salary in 1147/1735, at the second division of the Corps of the Bombardiers in the *inspection book* (*Yoklama Defteri*), (Primeminister Ottoman Archives (BOA) *Maliyeden Müdevver* (*M.MD*), n° 5941, 22). Although, according to (E. Ihsanoğlu, *OMLT*, vol. I, 180), he wrote in Turkish a treatise called *Risale fi'l-Mesaha* which was copied by Said Efendi, in fact *Risale fi'l-Mesaha* and *Fasl-ı Mesaha bilâ Alet* (in Arabic) were Said Efendi's own treatises. *Duvara Inhiraf-ı Saat-ı Nehâr Istihrac Eyleme Tariki* (Süleymaniye Library Esad Efendi n° 3704, 89-93) and *Medarât Vaz' Etmenin Tariki* (Süleymaniye Library Esad Efendi n° 3704, 94-95) are two known works of Abdullah Efendi which was narrated to Said Efendi orally.

2. M. Kaçar, " Osmanlı Devleti'nde Askeri Sahada Yenileşme Döneminin Başlangıcı ", *Osmanlı Bilimi Araştırmaları* (I), in F. Günergun (ed.), Istanbul, 1995, 209-225.

The sundials located on the walls of the mosques have a special religious importance. While constructing these sundials, the deviation of the meridian plane, perpendicular to the horizon, have to be determined correctly. Said Efendi gave a method to measure the correct deviation of the east-west direction relative to the meridian of the sundial. As seen in related figures, different from the sundials normally used by the Ottomans, where the drawing starts from 1 and ends at 6, his drawing starts from 6 o'clock in the morning and ends at 12 o'clock in the evening just like European clocks. His work also contains basic information about the construction of sundials and the ways of marking them on a wall — which has now probably been forgotten.

This work of Said Efendi shows a new perspective on the introduction of European science in the Ottoman world and its effects on Ottoman science. European science reached the Ottoman Empire by three main ways : Firstly, through translations, secondly, through visitors and finally by newly established institutions. We can say that this last way was through learning directly from the European scientists who worked in the service of the Ottoman Empire.

Let us see how Said Efendi draws a sundial deviating from the east-west direction. First, on the vertical sundial plane, by means of a setsquare a horizontal *AB* line (*hatt-ı ufki*), parallel to the horizon and a vertical *CD* meridian line (*hatt-ı nısfu'n-nehar*) perpendicular to this line are drawn (Figure 1).

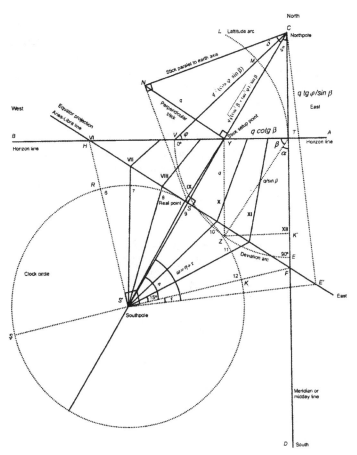

β = the angle of the wall with south direction α = 90°-β the angle of the wall with west-east direction

φ = latitude of the sundial (Istanbul $\varphi \cong 41°$) q = length of the stick perpendicular to the wall

Figure 1

At point *T*, where these two lines intersect, *T* is taken as center and *TV* as radius, and a deviation arc *VE* (*kavs-i inhiraf*) is drawn. Midday (*vakt-i zeval*) is determined with the aid of an " Indian circle " (*daire-i Hindiye*). The shadow of a stick in the center of the " Indian circle ", which is oriented according the geographic directions, at midday time falls along the north-south direction. At this moment, if the sun is located on the right side of a plain perpendicular to the wall, the deviation is towards east (*inhiraf-ı şarki*), and if the sun is located to the left, the deviation is towards the west (*inhiraf-ı garbi*) (in Figures 1 and 2 the deviation is towards east). The side of a quadrant is placed horizontally and tangently to the wall, in such a manner that the altitude arc of the quadrant will look towards the sun (Figure 2). At the same moment the shadow of a plumb line is made to fall on the center of the quadrant ; and with the help of the shadow passing through the center of the quadrant, the angle α between the west or east side of the wall and the south direction is read on the altitude arc and registered.

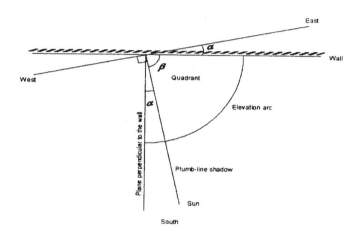

Figure 2

Hereafter, the deviation arc *VE* is drawn, by taking *T* as center and *TV* as the radius of a circle. The first quarter of the deviation arc *VE* (*ahad-ı rubu'l-esfe-leyn*) is scaled at 90°. On this arc, beginning from the horizon line, the deviation angle α and adjacent to it the angle between the wall and the south direction β = 90° - α is marked (Figure 1). From the point *Z* determined by this angle, the sinus line (*ceyb-i maksut*) *ZY* is drawn perpendicular to the horizon and parallel to the meridian line. The length of this line is a measure for the steepness of the latitude (*miktar-ı tul-ı amud*), and is equal to the length *q* of the stick. Also, from the above given point *Z*, a cosine line (*ceyb-i makus*) *ZK'* parallel to the horizon and perpendicular to the meridian is drawn and the center perpendicular to the meridian or the midday point (*bu'ud-ı merkez-i 'amud ez-nısfu'n-nehar*) *K'* is found. Then, the point *V* where the deviation arc intercepts the horizon as center and the radius equal to *VT*, a latitude arc *L* (*kavs-ı arz-ı beled*) of a quarter length is drawn. On this arc, beginning from the deviation arc center, an angle *MVT* equal to the φ latitude is marked by a pair of compasses. The marked M point is joined with *V* latitude arc center by a straight line and is extended to the meridian line. The intersection point *C* is the northpole (*kutb-ı şimal*) or the intersection point of all the shadow lines.

Thereafter, according to the deviation of the wall to east or west direction, the meridian will be at the east or west side of the line. The point *Y*, where the stick is set up (*merkez-i amud*), is joined to the northpole *C* and extended to the direction of the pole axis (*hatt-ı sath-ı nısfu'n-nehar*) and the line *CY* is obtained. From point *Y* where the *CY* pole axis intercepts the horizon, a perpendicular line is drawn. On it, a distance (*re's-i amud*) which is equivalent to the perpendicular latitude distance $NY = ZY = q$ is marked with the aid of a compass. This point *N* is united with the northpole *C*. By this method, a perpendicular pole shadow (*kutr-ı zıll-ı amud*), or the real length *NC* of the stick shadow (*tamam-ı zıll-ı menkus*), which is parallel to the pole axis and projected on the plane of the sundial, is obtained.

Next, from the end point of the stick *N* (*re's-i amud*), a line perpendicular to *NC* is drawn ; this line intersects the pole axis at point *S*, named the " real point " (*nokta-ı asl*). From the real point, a line perpendicular to the pole axis is drawn. This line intersects the horizon line at *H*. This line is extended in two directions and is called the Aries-Libra line *HSF*. Then, the distance *NS* between the shadow stick end N and real point *S* is carried, with the help of a compass *S'* = *NS* to the south *CYS* direction, and the southpole (*kutb-y cenup*) *S'* is obtained. Considering the southpole *S'* as center and *S'S* as radius, the clock circle (*daire-i fazl-ı dair*) is drawn. The southpole *S'* and the point F where the Aries-Libra line intersects the meridian line are joined as *S'F* and extended to form the radii of the circle *SK*. If this line is perpendicular to *S'H*, a line which connects the point *H* (intersection point of Aries-Libra and horizon line) with southpole *S'*, there is no mistake in the drawing of the sundial. Each quarter of the clock circle is divided into six equal parts of 15 degrees. These points are extended to the Aries-Libra line passing from the center *S'*. The points 6-12 obtained on the Aries-Libra line are extended to the horizon line and joined with the northpole, so that for each hour a shadow line is obtained[3].

---

3. Previous studies give detailed information about deviation determination and drawing vertical sundials. See Gazi Ahmet Muhtar Paşa, *Riayazu'l-Muhtar*, Cairo Bulak Press, Egypt H. 1303/1886, 76-77 and *Zeyl* (appendix) figure 34 ; Joseph Drecker, " Die Theorie der Sonnenuhren ", *Die Geschichte der Zeitmessung und der Uhren*, Band I, Lieferung E, Berlin, Leipzig, 1925, 51.

Calculations of the Sun orbit projections

On a wall surface, a longitude line called Aries-Libra orbit and a line perpendicular to it, the projection of the earth axis, are drawn (Figure 3). The point N, in which the lines intersect\ is taken as the center of a circle with an arbitrary radius and is divided into 360 parts. From the center N we take an interval NC as long as the stick length (Figure 1) parallel to the earth axis (*kutr-ı zıll-ı amud*). This point is called northpole C (*kutb-ı şimal*). Afterwards, beginning from the Aries-Libra orbit, on the four sides of the circle the 23°30″ points are marked and the center of the circle is connected to these points and extended.

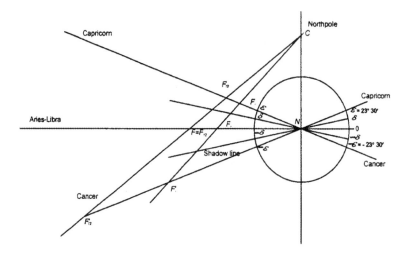

Figure 3

The upper ends of the x shaped figure determine the sun's Capricorn, and the lower ends the sun's Cancer orbit. Afterwards, the interval between the Aries-Libra orbit and the northpole at midday line $CF = CF_{12}$ is carried and marked with a compass, one leg at the northpole C and the other on the Aries-Libra line. If these points are connected, the intervals $(CF_{12} - CF_{12}')$ and $(CF_{12} - CF_{12}'')$ placed between the F_{12}' Capricorn and F_{12}'' Cancer lines, determine the declination of the sun, on the Capricorn and Cancer orbits. In the same way, for the other hours, the intervals between the F_i Aries-Libra orbit points and the northpole C are carried and marked with a compass. For the other hours of the sun's Capricorn and Cancer orbit, the shadow intervals between the points F_i' and F_i'' are obtained in the same way. Finally, the sun orbits for the declination angles $|\delta| < \varepsilon = 23°30'$ are calculated in the same manner and marked on the sundial surface. Probably, in order not to complicate the drawing these intervals are not shown in the original figures.

Conclusion

The Corps of Bombardiers established under the administration of a European general in the beginnings of the 18th century and forming a significant stage of Ottoman modernization was at the same time an institute where people from various cultures met, be them from the lands of the Ottoman Empire or from Europe. This contributed to the exchange of knowledge and know-how among various cultures. This work of Said Efendi is a good example for this fact.

Said Efendi's construction corresponds exactly to the drawing of vertical sundials deviating from the east-west direction. He produced this construction in the conventional style by making use of astronomy and spherical geometry. In his work, he provided the theoretical base of the method of rendering the degree of deviation from the wall plane. As distinct from conventional Ottoman sundials, he used some aspects of European ones. Just like the European sundials his construction started from 6 o'clock in the morning instead of 1 o'clock, which was the case with the Ottoman sundials.

APPENDIX

1 - The Construction of vertical sundials deviating from the east-west direction

Duvara Inhirâf-ı Saat-ı nehâr Istihrâc Eylemenin Tarikini Beyân Eder

Evvel, sath-ı duvara zât-ı müsellesle ufka müvâzî hatt-ı ufkî ihrâc ve ol hattı terbi' edip ona hatt-ı nısfu'n-nehâr deyip, ba'dehu tekātu'-ı hatteyn merkezinden bir dâi're-i tâm edip kavs-i inhirâf diyesin. Ba'dehu dâi're-i hindiye ile vakt-i zevali malûm edip duvara mukābele edip şems yemînde ise inhirâf şarkî solunda ise garbî olur. Rubu'-ı dâi'renin bir hattını duvara mümâss ede kavsü'l-irtifa şemse arka vere. Ba'dehu ol vakit bir şakullü haytı zıllı merkez-i rub'a rast gelince tahrîk ve o halde kavsü'l-irtifa'dan duvar cânibinden kaç derece hayt kat' eder ise ol kadar hatt-ı maşrık ve'l-mağribden duvarın inhirâfı olur. Anı hıfz edesin. Ba'dehu kavsü'l-inhirâf dâi'resinin 'ahad-ı rub'u'l-esfeleyni (Sat=90) mütesâviye taksîm ve ufuktan derecâtını resm edesin ve ufuktan inhirâf mikdârı derece sapıp hatt-ı ufka 'amûd ve nısf-ı nehâra muvazi bir ceyb-i mebsût ihrâc edip mikdâr-ı tûl-ı 'amûd diyesin ve yine derece-i mezbûrdan ufka müvâzî nısf-ı nehâra kāi'm bir ceyb-i ma'kûs ihrâc edip ve zıll-ı menkûs ve bu'd-ı merkez-i 'amûd ez nısf-ı nehâr diyesin. Ba'dehu kavsü'l-inhirâfın ufka tekātu' ettiği noktadan yine o kavsin nısf kutr bu'dunda râb'i-i a'laya bir rub' kavs çizesin ki buna kavs-i 'arz-ı beled derler ve kavs-ı inhirâftan 'arz-ı beled mikdârı derecede feth-i pergâr edip merkez-i kavsü'l-inhirâftan feth-i mezkûr mikdârı kavs-ı 'arz-ı belede bir nokta vaz' edesin kavs-ı 'arz-ı beledin merkezinden ol noktaya mârr ve ondan hatt-ı nısf-ı nehârın 'âlâsına mütecâviz bir hatt-ı müstakim ihrâc edesin ol hat hatt-ı nısf-ı nehârla mütekatı olduğu nokta kutb-ı şimâli duvar olur ve bu nokta mecmû'-ı sa'atin hutûtunun mültekāsı olup ol noktadan çekilir. Ba'dehu inhirâf-ı duvar şarkî ise hatt-ı nısf-ı nehârdan şarka, garbî ise garba mikdâr-ı bu'd-ı merkez-i 'amûd ufuk üzerine bir nokta vaz' edip merkez-i 'amûd diyesin ve kutb-ı şimâlden merkez-i 'amûda mârr bir hatt-ı gayr-i mütenâhi ihrâc edesin ki buna hatt-ı sath-ı nısfu'n-nehâr derler. Ba'dehu hatt-ı sath-ı nısf-ı nehârın ufukla tekātu'u ki merkez-i 'amûddur andan hatt-ı sath nısf-ı nehâra bir 'amûd-ı ulvî ihrâc edesin ba'dehu tûl-ı 'amûd kadar feth-i pergâr edip merkez-i 'amûd noktasından hatt-ı sath-ı nısf-ı nehâr üzerine ihrâc ettiğin 'amûddan tûl-ı 'amûd mikdarı fasl edesin ba'dehu kutb-ı şimâlle 'amûd-ı mefsûlün re'sini vasl edesin ki buna kutr-ı zıll-ı 'amûd ve tamâm-ı zıll-ı menkûs derler. Ba'dehu re'is-i 'amûddan kutr-ı zıll-ı 'amûd üzerine bir 'amûd ihrâc edesin ta hatt-y sath-ı nısf-ı nehârı tekātu edince, ol tekātu noktasına nokta-i asl diyesin ve nokta-i asıldan hatt-ı sath-ı nısf-ı nehâr üzerine canib-i ufka bir 'amûd ihrâc edesin ta hatt-ı ufku bir noktada tekātu ede ve bu 'amûdu hatt-ı nısf-ı nehâr cânibine gayr-i nihâye ihrâc edip medâr-ı haml ve'l-mizân diyesin. Ba'dehu nokta-i asılla re'is-i 'amûdun muvassıllı olan hat mikdarı hatt-ı sath-ı nısf-ı nehârdan esfel medâr-ı hamlde fasl edip nokta-ı fasla kutb-ı cenûb diyesin ve kutb-ı cenûbtan nokta-i asıla varınca nısfı kutr edip kutb-ı cenûb merkezinden bir dâi're-i tamâma resm edesin ve buna dâi're -i fazl dâir diyesin. Ve nokta-i cenûb ile medâr-ı hamlin hatt-ı nısfu'n-nehâr ile tekātu noktasını vasl edip dâi're-i fazl-ı dâi're varınca a'lel-istikame kutr edesin ve işbu kutra kaim a'lâye nısf kutr dahi ihrâc edesin ve ila-gayri'n-nihâye bu nısf kutru ihrâc edesin eğer medâr-ı hamlin ufk ile tekātu ettiği noktaya mârr olursa amel sahîhdir. Ba'dehu dâi're-i fazl-i dâi'rin muhitini a'le't-tesâvi taksîm yani her rub'unu altıya kısmet ve bu kısmette gayet ihtimâm edesin ba'dehu merkezinden bu altı kısmet noktalarına mârr altı hat ihrâc edesin ki medâr-ı hamlde bir nokta tekātu ede ba'dehu kutb-ı şimâlle ol tekātu noktalarını muvassıl olacak ufuktan ol noktalara birer hat ihrâc edesin ve her hat bir saat olur nısfu'n-nehâr i'tibârına göre ve's-selâm

Garbî   Şarkî

2 - Calculation of the Sun orbit projection

Medârat Vaz' Etmenin Tariki

Bir sath-ı duvar önüne bir hatt-ı müstakîm-i tûlî ihrâc ve medâr-ı haml ve'l-mizân diyesin ve ortasına bir 'amûd ihrâc ve merkez-i 'amûda bir dâi're resm ve dâi're-ŞS (360) mütesaviye kısmet ve kutr-ı zıll-ı 'amûd kadar ve bu 'amûddan fasl ve nokta-i fasla kutb-ı şimâl diyesin. Ba'dehu medâr-ı hamlden mizân ve dâi're-i mez-kûreden 23 30 mikdârı dört tarafına nişan edip merkez-i dâi'reden ol dört nişan dört nısf kutr ihrâc edesin ve ila gayri'n-nihâye çekesiz. Hakezâ X hat-teyn-i uluvviyine medâr-ı cedi süfliyinine medâr-ı sertan diyesin. Ba'dehu münharifede nokta-i şimâlle medâr-ı haml ve mizân hatt-ı nısfu'n-nehâra tekâtu ettiği nokta mabeyni kadar feth-i pergâr edip vaz'-ı medâr mıstarasında kutb-ı şimâlden medâr-ı haml ve mizân üzerine nişân edip ol nişâna marr kutb-ı şimâlden hatt-ı müstakîm ihrâc ve bu hattan medâr-ı haml ve mizân ve medâr-ı cedi mabeyni mün-harifede hatt-ı nısf-ı nehâr üzerinde medâr-ı haml ile medâr-ı cedi(ni)n mabeynidir ve bu hattan medâr-ı haml ve medâr-ı sertan beyni mün-harifede hatt-ı nısf-ı nehârda medâr-ı haml ve medâr-ı sertan beynidir. Ve kezalik sâir sâ'atin dahi münharife kutb-ı şimâlden medâr-ı haml ile saat tekātı' ettiği nokta mikdâr feth-i pergâr ve mıstarada kutb-ı şimâlîden medâr-ı haml ve mizân üzerine nişân ve mabeyni vasıl-ı hatlar ihrâc ve her sâ'atin medâr-ı haml ve medâr-ı ciddi ve medâr-ı sertanın bulup münharife üzerlerinde nişân ve beynlerini vasl edesin. Medârât hasıl olur.

Ahaztü (1148) ve ketebtü hâzihi'r-risâlete mîn lisân-ı Abdullâh Efendi el-Muzafferî el- Bosnavî fî mahmiyeti Üsküdar fî meştây-ı Humbaracıyân 'alâ sahifetihi. li ennehu yürîdu'z-zihâb gadihi 'ilâ vatânihi. Sümme nakale ilâ hunâ min Mustafa b. Ibrâhîm el-Kâtib fî'l-Ocağ-ı mezkûre. Sümme kabeltu bi-aslî'n-nüshati ve sahhahtü mehmâ kadertu. Ve ene'l-fakîr ilâ keremi Rabbî vâs'i'i'l-mağfireti Mehmed Said b. el-Müftî fî Beyşehrî el-Hacc Mahmûd Efendi b. el-Hacc Hasan Efendi b. Ahmed el-Haddâd fî yevmi'l-Cum'a es-sâbi' 'aşer min Muharremi'l-haram li sene hamsîn ve mi'e ve elf (17 Muharrem 1150/17 May 1737)

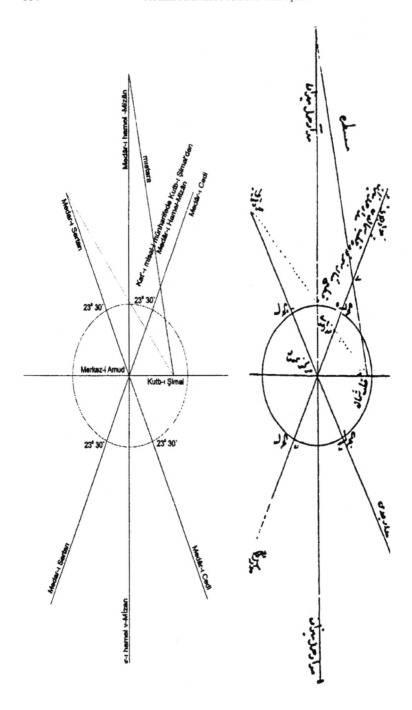

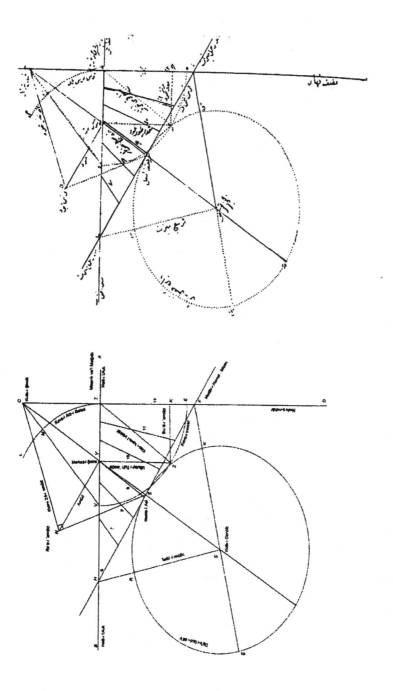

SCIENTIFIC PRACTICE, PATRONAGE, SALONS, AND ENTERPRISE IN EIGHTEENTH CENTURY CAIRO : EXAMINATION OF AL-GABARTĪ'S HISTORY OF EGYPT

Rainer BRÖMER

ISLAMIC CIVILISATION IN THE EIGHTEENTH CENTURY AD

The eighteenth century is perhaps the least studied period in the history of the Islamic world, certainly in the field of science and learning. At the same time, it is a most relevant epoch for the general history of the Middle East where the stage is set for massive military and social interventions by European powers in the following century. The consequences of European[1] imperialism and colonialism for the spread of Western ideas about science and education have been extensively studied. Muḥammad ʿAlī's efforts to appropriate particularly French military and administrative technology, hiring European teachers and sending students across the sea in order to make sure that the events of 1798-1801, the abortive French occupation of Egypt, would never be repeated, are well known to any student of modern Islamic civilisation — taken by some as the ultimate defilation of Islam, by others as a laudable effort to reach modernity but falling short of demolishing the old thoroughly enough[2]. Rifāʿa Rāfiʿ al-Ṭahṭāwī is a household name in these debates : the imam, translator, author, and educator, whose diary of his stay with one of Muḥammad ʿAlī's student missions in Paris was among the first Arabic books ever written with the perspective of printing in mind.

However, we must not forget that the events of 1798 didn't happen in a historic vacuum, neither from the Egyptian side nor from the French. The intimate connection between French Enlightenment thought and the political and intel-

1. And soon American, too ; see Samir Khalaf, " New England Puritanism and Liberal Education in the Middle East : The American University of Beirut as a Cultural Transplant ", in Şerif Mardın (ed.), *Cultural Transitions in the Middle East*, Leiden, Brill, 1994, 50-85.

2. The latter view was proposed e.g. by Sabry Hafez, " The Novel, Politics and Islam ", *Haydar Haydar's Banquet for Seaweed*, New Left Review, 5 (Sep./Oct. 2000), 117-141, esp. 119.

lectual preparation for Napoleon's adventure has also been studied extensively[3] : No wonder, as the Age of Enlightenment has always attracted the attention of European historians. Geostrategic concerns, Egypt's position on the route to India, as well as the symbolic meaning of the orient (*ex oriente lux*) seem to have stimulated the fantasies of French officials seeking to get the quarrelsome Corsican general out of the way in the troubled days of the young republic after the Terror.

When the French troops marched up the Nile without meeting with any noticeable resistance, the inhabitants of Cairo fell into a state of deep shock, if we are to believe the Egyptian chronicler ʿAbd al-Raḥman al-Gabartī (1167/ 1753-4 - ca. 1240/1825[4]) who eyewitnessed the disturbing events. Not that things had been in the best of states in Egypt previously. Gabartī's large chronicle, spanning mainly the eighteenth and first fifth of the nineteenth century of our time, reports of an equal amount of 'murder, mayhem, pillage and plunder' as Miḫāʾīl Mishāqa's later Syrian history[5]. But the centre of power, though distant, had been the Sublime Porte, the sultan being at the same time recognised as the caliph, spiritual leader of Islam, whereas Napoleon's bogus Islamic swindle was all too easily understood for what it was (not only by the ʿulamāʾ) : a transparent disguise for the onslaught of the infidels on a central land of Islam. In a similar vein, the learned men of Cairo were unimpressed with the demonstrations of Western scientific 'superiority' at the Institut d'Egypte that Napoleon had established as a base for the exploration of the 'cradle of civilisation' on the Nile, but which also was to serve in an attempt to win over at least some of the notables as collaborators for the occupation regime[6], while most seem to have remained sceptical, exhibiting a strong sense of awareness of their own autochthonous cultural resources (III, 34)[7].

It is quite revealing that Western readers are only familiar with the Napoleonic episode from Gabartī's chronicle and, similarly, with Ṭahṭāwī's Description of Paris, equally reserved towards the infidels' achievements, while both of these writings only represent a tiny fraction of the literary production of each of them.

    3. Henry Laurens, " Les origines intellectuelles de l'expédition d'Égypte ", *L'Orientalisme Islamisant en France (1698-1798)*, Istanbul, Paris, ISIS, 1987 ; see also Patrice Bret (éd.), *L'expédition d'Égypte, une entreprise des Lumières 1799-1801*, Paris, Technique & Documentation, 1999.
    4. The date of his death is uncertain ; *cf.* Thomas Philipp & Guido Schwald, *ʿAbd al-Raḥman al-Jabartī's History of Egypt, Guide*, Stuttgart, Steiner, 1994, 5 and fn. 14.
    5. Mikhâyil Mishâqa, *Murder, Mayhem, Pillage, and Plunder*, Albany, SUNY Press, 1988. The sensationalist title was chosen by the editor of the English translation.
    6. *Cf.* the fate of Šaiḫ Bakrī and his daughter, John W. Livingston, " Shaykh Bakrī and Bonaparte ", *Studia Islamica*, 80 (1994), 125-143.
    7. On more nuanced reactions, *cf.* J.W. Livingston, " Shaykhs Jabartī and ʿAṭṭār : Islamic Reaction and Response to Western Science in Egypt ", *Der Islam*, 74 (1997), 92-106.

THE PROBLEM OF THE SOURCES

The question this paper wishes to address regards the intellectual situation in Cairo before the French invasion which apparently had not been anticipated (at least we don't find any signs of perceived threads from any infidel power) and can therefore not be used as any kind of culmination point of the cultural evolution during the eighteenth century. If we want to understand eighteenth-century culture in Egypt in its own terms, we get into difficult territory. For one thing, the availability of firsthand accounts from the region seems to be limited, although (as I hope to show later on) the lack of attention to the extant material might be an equally serious problem as the loss of sources. In the case of Cairo, we are in the lucky situation to possess al-Gabartī's History, which of course we have to approach with due caution in order to assess the impact of the author's personal preferences on the presentation of his narrative. We will see one example of this potential bias further down, which might even be advantageous in as far as it coincides with our interest in history of scientific practice.

But there is more in our favour : David King's inestimable work that produced a Catalogue of Scientific Papers held in the National Library in Cairo[8] shows that a large number of writings from Gabartī's protagonists have actually survived *in situ*, and from the diligent compilation in Ihsanoğlu's histories of Ottoman astronomical and mathematical literature we find numerous manuscripts of these same authors that made their way into libraries around the world.

The starting point for this paper, therefore, is a programmatic perspective on the possibility of pursuing a case study on the history of science in eighteenth century Cairo as an example for what one might call a scientific community in one of the major secondary centres in the Ottoman Empire, a place that was still renowned among the learned for its learning. There is one feature that will emerge only tangentially, and that is the possibility that intellectual exchange, so much favoured across the Muslim world, may have included the crossing of the Mediterranean as well. Certainly, diplomatic and economic history has well documented the intellectual aspects of Western presence in Levantine ports. After all, local merchants were dependent on overseas markets and therefore clearly interested in the changes in European markets that affected their trade. In how far it is admissible to consider the Arabic (and Turkish, for that matter) notables as part of one Euro-Mediterranean civilisation in the Age of Enlight-

---

8. David A. King, *A Catalogue of the Scientific Manuscripts in the Egyptian National Library*, Egyptian Book Org., 1981, 2 vols ; see also *id. A Survey of the Scientific Manuscripts in the Egyptian National Library*, Winona Lake, Eisenbrauns, 1986 (American Research Center in Egypt Catalogs, 5).

enment is an extremely problematic issue — as especially German reading scholars of modern Middle Eastern history will have realised. To be sure, this paper is not going to address the question if there was an Islamic Enlightenment raised by Reinhard Schulze some years ago[9]. The ongoing polemic between Schulze and Radtke[10] is focused on intricate philological issues far beyond the reach of my linguistic skills, and the possible outcome will adjudicate over a fascinating, but relatively abstract concept of Enlightenment structures in literary expression[11], whereas my aim is far more modest, looking at a corpus of far less poetically aspiring text that can on all accounts be considered scientific, certainly in the tradition of classical Arabic science, but potentially being more than simply epigonous works or re-calculations of medieval zijes (azyāǧ). This reflection ties into a broader programme of research, in opposition to the traditional decline thesis, devoted to scientific activity after the " classical " period in the Islamic civilisation which, in the particular case of the Ottoman Empire, Ihsanoǧlu has specified as " Ottoman Science "[12]. At the same time, we have to acknowledge that our protagonists did have personal contacts with travellers from Europe[13] and read foreign texts[14], which supports Ihsanoǧlu's contention that intellectual exchange did play a significant role[15].

DOES AL-GABARTĪ DESCRIBE A SCIENTIFIC COMMUNITY IN EIGHTEENTH-CENTURY CAIRO ?

From Gabartī's chronicle we can glean quite detailed information about the social interaction between notables and scholars, including more formal settings like salons (maǧālis) and structures of patronage which can, with due caution, be compared to institutions of early modern Western science. The main body of this paper will be dedicated to the presentation of some examples for different types of scientific activity. The question to be answered by further

9. Reinhard Schulze, " Was ist die islamische Aufklärung ? ", *Die Welt des Islams*, 36 (1996), 276-325.

10. Bernd Radtke, *Autochthone islamische Aufklärung im 18. Jahrhundert*, Utrecht, Houtsma Stichting, 2000. It seems to be part of the polemic when Radtke does not place a question mark at the end of the title, since he vehemently rejects Schulze's claims.

11. Before this latest debate, the phenomenon of a literary revival in Arabic culture had been studied by Usama 'Anuti and Peter Gran ; see Usāma 'Ānūtī, *al-Ḥaraka al-adabīya fī bilād al-Šām ḫilāla 'l-qarn al-ṯāmin 'ašar*, Bairūt, Univ. Libanaise, 1971, esp. 237-265 ; Peter Gran, " Islamic Roots of Capitalism ", *Egypt, 1760-1840*, Austin, London, UTxP, 1979.

12. E.g. Ekmeleddin Ihsanoǧlu, " Some remarks on Ottoman Science and its relation with European Science & Technology ", *Journal of the Japan-Netherlands Institute*, 3 (1991), 45-73. For a more recent overview, *id.* & Feza Günergun (eds), *Science in Islamic Civilisation*, Istanbul, IRCICA, 2000.

13. See below under Ḥasan al-Gabartī.

14. An example being al-'Aṭṭār's study of European medical texts during his stay in Istanbul, Gran (1979), 104f.

15. E. Ihsanoǧlu (ed.), *Transfer of Modern Science and Technology to the Muslim World*, Istanbul, 1992.

studies based on the contemporary manuscripts is the following : What was the actual intellectual content of these activities ? Was it, as is often alleged, a mere continuation and adaptation of ancient knowledge, such as the geographical conversion and updating of astronomical tables from Samarqand, or did the authors express any intention of improvement and novelty, however concealed to avoid accusations of illicit innovations (*bidaᶜ*) ? Should we rather compare the occupations of Riḍwān al-Falakī or Yūsuf al-Kilārǧi to the widespread trade of almanac makers in Europe at the time ? But then, of course, there was no boundary of any kind between making almanacs and contributing to the process of science in the century before the scientist as a specific identity came into being, which suggests that with respect to al-Ḫawānakī or the chronicler's father, Ḥasan al-Gabartī, the question might be as badly framed.

But let us now turn our view on ᶜAbd al-Raḥman al-Gabartī's major work, *ᶜAǧāʾib al-āṯ ār fī ʾt-tarāǧim wa-l-aḫbār*, written towards the end of the author's life in the 1820s, printed at Būlāq in four volumes in 1879/80 and only fully translated into English in 1994[16] — while a French edition had been published in the 1890s in Cairo.

For the period before his and his father's lifetime, Gabartī explicitly relies on Murtaḍā al-Zabīdī's monumental Tāǧ al-ᶜarūs[17], and the chapters are cumulative over several years, whereas in the later parts, every year (eventually each month) is documented in a separate chapter, each subdivided in a more or less lengthy description of the major events and an extensive number of necrologies of notables who died in that year, including collections of other writers' obituaries. Most information on scientific pursuits is included in the necrologies, while only exceptionally issues of learning made it into the general sections of the annual entries which focus on political events and crime stories.

While this paper will only examine a few characteristic examples, it must be said that there are quite a few more, so that if we keep in mind that we are looking at but one, though important, city over the period of one hundred years, it is quite striking to note the relevant number of individuals who are

16. ᶜAbd ar-Rahman al-Jabarti, *History of Egypt*, vol. 2, in Thomas Philipp & Moshe Perlmann (eds), Stuttgart, Steiner, 1994, 4 vols. References in the present article use the volume and page numbers of the Būlāq edition which are preserved in the translation. For crosschecks of the translation, an undated (ca. 1970 ?) edition of *Tārīḫ ᶜAǧāʾib al-Āṯ ār fī ʾl-tarāǧim wa-l-aḫbār li-l-ᶜallāma al-šaiḫ ᶜAbd al-Raḥman al-Gabartī, Dār al-Ǧīl*, Bairūt, 3 vols, has been used, although it has been severely abridged (omitting mainly the collections of other authors' necrologies, but also the iǧāzāt which provide invaluable information about the range of learning available to our protagonists). Quotes from this edition are preceded by " Bairūt ".

17. On Zabīdī, see Stefan Reichmuth, " Murtaḍā az-Zabīdī (d. 1791) in Biographical and Autobiographical Accounts. Glimpses of Islamic Scholarship in the 18th Century ", *Die Welt des Islams* 39 (1999), 64-102.

mentioned as being involved in some sort of scientific activity. What will
emerge is a (necessarily sketchy) prosopography of a distinct group of individuals tied together by a common interest in scientific knowledge, though this
should not obscure the observation that — even in the account of the scientifically inclined chronicler — the main focus of the *maǧālis* was literary, with a
strong emphasis on *ṣūfī* spirituality.

Two central figures for the astronomical and mathematical sciences can be
found who, in subsequent generations, create certain foci of learning, observation, and mechanical technology : Riḍwān Efendī al-Falakī at the turn of the
18th century and Ḥasan al-Gabartī, not incidentally the father of our historian,
in the middle half of the century. It is quite clear that ʿAbd al-Raḥman, the son
of Ḥasan, devotes exceptionally large space to the narration of his father's life
and deeds, and this might be one reason for the striking bias towards those disciplines the father was most involved in. In contrast, we learn very little for
instance about medicine[18].

ASTRONOMICAL TABLES AND INSTRUMENTS : RIḌWĀN AL-FALAKĪ AND HIS PATRON ḤASAN EFENDĪ

The most recurrent name in the eighteenth century in Cairo seems to be the
one of Riḍwān Efendī al-Falakī, the astronomer (d. 1122/1710). Ihsanoğlu
calls him the perhaps greatest Egyptian astronomer of his time and provides a
list of seventeen writings most of which survive in multiple, up to eighteen
copies — which is more than many printed books in Europe of the time[19]. Not
much seems to be known about Riḍwān's personal background. Al-Gabartī (I,
74) calls him " the honoured scholar " but does not provide any genealogy, nor
kunya ; however, he presents him to his readers as a prolific author : " In his
own hand he wrote more than a camel load of manuscripts and tables ".
Remarkably, in this context, is the formulation that Gabartī uses to describe
Riḍwān's *zij*, " which he composed on the pattern of al-Durr al-Yatīm by Ibn
al-Majdi according to the principles of the new astronomical system of al-

18. On the one hand, as Peter Gran (*cit.*) documented, there actually seems to have been a crisis
in the medical system in Cairo at the time, and medical care for the Arab population tended to be
neglected by the Turkish elites ; however, as we know from the biography of ʿAbd ar-Raḥman's
contemporary al-ʿAṭṭār, medical science still was a major issue at their time and by no means
ignored — but apparently it was not an occupation of Ḥasan al-Gabartī and therefore outside the
chronicler's focus. As al-ʿAṭṭār survived ʿAbd al-Raḥman by a few years, he obviously is not
included in Gabartī's chronicle.
19. " en büyük astronom diyebileceğimiz ", E. Ihsanoğlu *et al.*, *Osmanlı Astronomi Literatürü
Tarihi*, 246, Istanbul, IRCICA, 1997, 377 (henceforth : *OALT*), listing a number of biographical references.

Samarqandī [*sic*] (I, 74) "[20], which raises the question for how many centuries a system can have been established without losing its novelty[21].

But in contrast to what we might expect from Gabartī's assertion that Riḍwān lived " secluded from human society, devoted to his own affairs ", he actually must have been quite a central figure in the intellectual networks of Cairo, since a number of scholars are introduced as his companions or students. Riḍwān was mainly working on astronomical calculations on the basis of the fifteenth century observations which Uluġ Beg had compiled in his zij. But what interests us here is that Riḍwān, whatever his own social background may have been, was supported by the patronage of an amīr, Ḥasan Efendī ad-Damurdāī, who had been given the office of Baş Kalfa of the ruzname by Ibrāhīm Paşa in 1694/95, apparently for his reputation as a skilful tax farmer (I, 114).

Ḥasan Efendī himself displayed a remarkable interest in astronomy[22], and he was willing to spend considerable amounts on his hobby. Around the year 1700, Ḥasan gave Riḍwān great sums of money to produce a series of astronomical spheres in gilded brass and other astronomical instruments. Gabartī mentions at the time of his writing, over a hundred years later, that " some of these instruments remain in Cairo and in other places to this day. They are engraved with his [Ḥasan's] name and with that of Riḍwān Efendī's ", (I, 115) which adds an interesting facet to this patronage relationship.

But Ḥasan also encouraged one of his servants, the mamlūk Yūsuf al-Kilārġī (d. 1153/1740), to engage in similar pursuits. Gabartī particularly emphasises Kilārġī's " excellent book on vertical sundials ; in it he collected the known research of former astronomers and added new discoveries concerning precise locations, orbits, and figures ". Isn't this quite a neat description of a positivist scientific programme : collect known data and add new discoveries ? Far from being rhetorical, this question should be answered by studying his writings, of which at least eight works survive, only one of which in three copies[23].

Another prolific writer of things astronomical in the first half of the eighteenth century is associated with Riḍwān and Yūsuf al-Kilārġī, namely

---

20. " alladī ḥarrarahu ʿalā ṭarīq al-durr al-yatīm li-bn al-Maġdī ʿalā uṣūl al-raṣad al-ġadīd al-samarqandī " (Bairūt, I, 130).

21. Gabartī repeats the exact same wording " al-raṣad al-ġadīd al-samarqandīi " in al-Ḥawānakī's necrology and elsewhere, while the translators' understanding varies, see below. The reference is to an astronomer of the period of Uluġ Beg's observatory in Samarqand in the first half of the fifteenth century. At the risk of appearing pedantic, is must be said that the most appropriate interpretation is needed for discussing the question concerning scientific innovation.

22. Bairūt (I, 178) has " art " (" fann ").

23. *OALT,* n° 275, 412-415, listing a number of biographical references.

Ramaḍān al-Ḥawānakī[24] (d. 1158/1745), of whom at least 25 texts have been preserved. Al-Ḥawānakī was also a close friend of our chronicler's father Ḥasan, and he owned some property which at his death he entrusted to Ḥasan al-Gabartī (I, 163).

So we find a group of people which are among other interests committed to astronomical observation, instrument making, and calculation of almanacs — useful pursuits for keeping religious and secular order, but the chronicle makes it clear that to a large extent personal enthusiasm prevails over practical requirements. Ḥasan Efendī's generous sponsorship not only supported the intellectual centre of the circle, Riḍwān, but even Ḥasan's mamlūk Yūsuf al-Kilārğī was put in a position where he could not only make observations and write about them, but according to ʿAbd ar-Raḥman he even had the resources to collect instruments and rare books. At the same time, other members of the circle had independent means, and thus their activities continued beyond the death of the first patron.

NOTABLE, MERCHANT, GENTLEMAN : ḤASAN AL-GABARTĪ'S SCIENCE AND CRAFTSMANSHIP

In 1748 a new governor arrived in Egypt, Aḥmad Paşa, who had come with high intellectual expectations. Our chronicler extensively illustrates the circumstances that significantly enhanced the glory of his father, but even considering his personal interestedness, the episode still provides valuable insight in the contemporary attitudes towards scientific learning. (I, 186-8).

Shortly after his arrival, Aḥmad Paşa contacted the ãiḫ al-Azhar, ʿAbd Allah al-ūbrāwī, trying to engage him in a conversation on mathematics from which the latter shied away. When the Paşa expressed his disappointment (" In Istanbul ", said the pasha, " Cairo has the reputation of being the fountain of virtues and sciences. I was therefore eager to come here ; but when I came, I discovered that, as the proverb says, 'Hearing about al-Muʾayyadī is better than seeing him.' "), ūbrāwī in his embarrassment admitted that al-Azhar was not the main domain of this kind of learning. Asked by Aḥmad Paşa where those knowledgeable in these sciences could be found, the ãiḫ significantly replied : " They are found in their homes, and people visit them there ", referring the governor to none other than Ḥasan al-Gabartī (d. 1188/1774)[25].

---

24. *OALT*, n° 279, 418-426. The spelling of the name varies ; Bairūt ed., 241f., has Ḫānakī (Persian pronunciation ?).

25. *OALT*, n° 314, 472-479 ; Ihsanoğlu *et al.*, *Osmanlı Matematik Literatürü Tarihi*, Istanbul, IRCICA, 1999 (henceforth : *OMLT*) n° 219, 219-221.

Here we have a precise acknowledgement that the private homes of (supposedly well-to-do[26]) individuals were the location for scientific interest, meeting in *maǧālis* rather than in *madāris* which were dealing with the transmission of canonic knowledge — a situation which quite neatly reflects the role of European universities prior to the mid-eighteenth century.

Back to the ǎiḫ al-Azhar who thus had saved the day : Ḥasan al-Gabartī would meet Aḥmad Paşa twice a week, studying mathematical works and later the drawing of sundials and other instruments.

Let's give back the word to our chronicler :

> Whenever the late Shaykh 'Abdallah al-Shubrāwī met my late father, he would say to him, 'May God cover your faults, even as you have covered over ours before this governor ! Were it not for you, he would have considered us all donkeys. God have mercy on us all.' "(I, 188).

Se non ĭ vero, ĭ ben trovato ; at any rate, from the perspective of the hero's son (who had not yet been born at the time of the visit) a lifetime later, this episode suggests a series of interesting observations about the institutional setting of scientific debate, the separation from the religious learning of the *madāris*, and the reputation Cairo still enjoyed, at least in the Ottoman Empire, but even beyond, as in 1159/1746-7, according to our chronicler,

> some European students[27] came to [Ḥasan al-Gabartī] and studied geometry. They gave him gifts of valuable objects and instruments of European manufacture. When they returned to their country, they published what they had learned, raising the level of such learning there from potentiality to actuality[28], and designing devices such as windmills, machines for lifting weights, pumping water, etc. (I, 397).

It would be interesting to track down these visitors, which to the best of my knowledge no-one has as yet succeeded to do. With all due reserve to the ulterior eulogy to his venerated father's reputation, it is at least imaginable that travellers in the mid-eighteenth century were as interested in Egyptian engineering (which *handasa* might as well refer to as to geometry, given the examples mentioned by 'Abd al-Raḥman) as were Napoleon's scholars some forty years later[29].

However, if we look at the context of the passage just quoted, we find another important aspect of scientific development, namely, the close associa-

---

26. Another host, Muḥammad al-Damurdāšī (d. 1178/1764-5), is described as " wealthy and prominent " (I, 265).

27. ṭullāb min al-ifranǧ (Bairūt, I, 461).

28. aḫraǧūhu min al-quwwa ilā 'l-fi'l (*ibid.*).

29. Gran (1979, 172) mentions that Napoleon charged his scholars to study Egyptian mills.

tion between bourgeois patrons and artisans : The mentioning of Frankish engineers follows a long list (the order seems significant) of valuable books Ḥasan al-Gabartī had purchased, then astronomical instruments (including the brass spheres that Riḍwān al-Falakī had designed for Ḥasan Efendī and which upon the latter's death were bought by al-Gabartī), in the next place " the tools of most trades "[30], followed by a list of " master craftsmen "[31], including two clockmakers, a locksmith, a cutler (Sākakīnī), and a kind of pharmacist. Finally, we learn that Ḥasan " devoted himself to craftsmanship " and demonstrated his skills to his servants who would then supervise the final execution of the works : " After his students had become proficient, my father stopped working at the craft and sent (later) students to them. (I, 398) " : It appears that improvement and teaching, but not the regular practice of mechanical trades was worthy of a gentleman[32].

But when important issues were at stake, the notable could well take up the tools again : a few years later, in 1172/1758-9, it was discovered that trade in Cairo was suffering from significantly unreliable weighing devices, and Ḥasan himself, in cooperation with manufacturers and operators, " calibrated them exactly according to the principles of applied science and geometry[33], spending his own money on the project out of a sense of religious obligation " (I, 399).

SCIENCE AS A LIVELIHOOD : FULL-TIME OR SIDELINE

An interesting person for the purposes of our survey, quite the opposite of the gentlemanly Gabartī, is āih Ḥusain al-Maḥalli al-āfi'ī (d. 1170/1756)[34] who apparently was able to make a living out of science, something our chronicler didn't approve of :

> He charged his students a fee and behaved arrogantly with whoever came seeking to study under him. He would agree to teach the works in question only after protracted haggling, saying : 'I don't sell knowledge cheaply'. He had a shop near the gate of al-Azhar, where he made a living selling almanacs for the computation of prayer times, various other books, and by copying. (I, 219f.).

Our chronicler apparently was more inclined towards the patronage system that allowed generous dispensation of knowledge, rather than the business of science : His own father, Ḥasan,

---

30. adawāt ġālib al-ṣanā'i' (Bairūt, I, 461).

31. kull mutqan wa-'ārif fī ṣinā'atihi (ibid.).

32. We might compare this attitude with the division of labour maintained e.g. in the early days of the Royal Society in London some eighty years earlier, cf. Steven Shapin, " The House of Experiment in Seventeenth-Century England ", ISIS, 79 (1988), 373-404.

33. al-'ilm al-'amalī wa-l-wad' al-handasī (Bairūt, I, 463).

34. OMLT, n° 128, 204-7.

was involved in trade — buying and selling, partnerships, speculation, and exchange. His grandmother was a woman of wealth who owned property and real estate. She placed a number of buildings into a *waqf* for [him] (I, 391),

which did not stop him from selling for a fortune the fur Aḥmad Paşa had given him as a remuneration for his efforts of teaching him (I, 187).

The last example presents a combination of scholarly and mundane activities, that of the tailor āiḫ Mustafa al-Ḥaiyāṭ who studied astronomy, among others, with Riḍwān Efendī al-Falakī, with al-Ḥawānakī, and with Ḥasan al-Gabartī. While continuing to oversee his tailor shop, he produced annual almanacs which, important for Egypt, included Coptic and Jewish dates — an interesting example of concern across the *milletler*.

However, when the Wafā'īya āiḫ[35] Aḥmad b. Wafā' desired to have an updated *zij*, he kept al-Ḥaiyāṭ's sustenance and that of his family for an extended period and rewarded him so lavishly that he could live on this account for months after the completion of the desired work. A warning regarding the use of the English translation of Gabartī's chronicle : as mentioned above, every single occurrence of " al-raṣad al-ǧadīd " is rendered differently, in Wafā''s case as " according to the most up-to-date observations ", which suggest empirical work (*arṣād ǧadīdat al-waqt*, if that makes any sense), whereas the phrase " baʿda 'l-raṣad al-ǧadīd ilā tārīḫ waqtihi ", corresponding to the preceding use of the term " al-raṣad al-ǧadīd ", suggests a (probably mathematical) conversion of the Samarqandī data, " according to the new *raṣad* [*i.e.*, the three-hundred year old Uluġ Beg tables] for the current date ", which unfortunately does not support the assumption that our eighteenth-century astronomers did any observational work. This emendation conforms to the translator's correct explanation in a footnote to al-Ḥaiyāṭ's necrology (1203/1788-9) where he cites the Persian title of the " Zīj-i jadīdī suḳṭānī [*sic*] by Ulugh Beg al-Samarqandī " (II, 181)[36].

CONCLUSION

If we want to use deliberately anachronistic and anatopic expressions, we might say that these examples selected from ʿAbd al-Raḥman al-Gabartī's chronicle of eighteenth-century Cairo represent patterns of gentleman science including interest in craftsmanship, patronage by a governor or by wealthy

---

35. The role of ṣūfī-orders like the Wafā'īya in the intellectual life in eighteenth century Cairo should not be neglected anyway, as Peter Gran (*cit.*) has shown in his account of the — mainly literary — *maǧālis*.

36. Actually, the translators change across the voluminous work, as Philipp (1994) documents in the *Guide*, 11.

individuals, the trade of almanac making, and the sale of knowledge as a small business (the latter being least well received).

Far from claiming any deeper significance for these analogies, the intention of sketching them is to draw attention to the complexities of scholarly pursuits in an Ottoman province at the time when we used to assume that there was nothing worth noting going on altogether.

The sheer quantity of accounts Gabartī provides in his chronicle[37] testifies for the non-irrelevance of science in pre-Napoleonic Egypt, even if we have to suspect that 'Abd al-Raḥman's commitment to the glorification of his father's intellectual achievements may have induced the chronicler to emphasise analogous aspects of other notables' lives, too, which are not prominent in other conventional *tarāǧim* dealing more with poetry and politics.

Starting from these observations, subsequent work on the so far unexamined sources will undoubtedly yield deeper insights into the history of the " dark " eighteenth century, in particular in the assessment of the intrinsic nature of the scientific interests and activities, the extent to which the new writings were mainly adaptations of classical works or contained elements of innovation and change. Close examination will be required, since for reasons of legitimation we cannot expect *bida'* being paraded ostentatiously.

This study will help us to challenge the decline thesis (according to which we shouldn't have found any sources at all), to qualify the " Islamic Enlightenment " thesis (which in our field would suggest innovative approaches to the classical material), and finally to reconstruct the intellectual stage onto which first Napoleon's scholarly retinue and then Muḥammad 'Alī's military instructors burst, creating a complex amalgam full of cultural tensions rather than simply introducing wholesale a new system of scientific thought into an empty space.

APPENDIX

List of scholars who died in Cairo during the eighteenth century, portrayed by al-Gabartī and included in *OALT/OMLT*, respectively[38].

al-'Uǧaimī, Abu 'l-Baqā' Ḥasan b. 'Alī b. Yaḥyā (1113/1702) ı, 69 [70 ?], *OMLT*, 101.

---

37. As this paper only discusses selected examples, the appendix provides a more complete list of scholars portrayed by Gabartī and included in *OALT/OMLT*.

38. Spelling of the names from *OALT/OMLT* converted to Rome transliteration, followed by year of death, page numbers from Būlāq edition of Gabartī's chronicle, number(s) in *OALT/OMLT*.

al-Bannāʾ, Aḥmad b. Muḥammad b. Aḥmad b. ʿAbd al-Ġānī al-Dimyāṭī al-āfiʿī (1117/1705) I, 89, *OALT*, 237.

Riḍwān b. ʿAbd Allah al-Razzāz al-Falakī (1123/1711) I, 74, *OALT*, 246.

al-Budairī, Muḥammad b. Muḥ. b. Muḥ. b. Aḥmad al-B. al-Ḥusainī al-Dimyāṭī al-Aʿarī al-āfiʿī Ibn al-Maiyik (1140/1728) I, 88, *OALT*, 261.

ʿAbd al-Ġānī al-Nābulusī, b. Ismāʿīl b. ʿAbd al-Ġānī b. Ismāʿīl ... al-N. al-Dimaqī (1143/1731) I, 154, *OALT*, 263.

Aḥmad al-Dairabī, Abu ʾl-ʿAbbās A. b. ʿUmar al-D. al-āfiʿī al-Azharī (1151/1738) I, 161 *OMLT*, 118.

al-Kilārğī [Turkish : kilerci], Gamāl al-Dīn Yūsuf b. Yūsuf al-Ḥalabī al-Maḥallī al-āfiʿī al-K. al-Falakī (1153/1153/1740) I, 164, *OALT*, 275.

Ramaḍān al-H̱(aw)ānakī, b. Ṣāliḥ b. ʿUmar b. Ḥiğāzī b. ʿUmar al-Safṭī al-H̱. (1158/1745) I, 162, *OALT*, 279.

Ḥusain al-Maḥallī, b. Muḥammad al-M. al-Azharī (1170/1756) I, 219, *OMLT*, 128.

al-Ḥifnī, āms al-Dīn Muḥammad b. Salīm al-Ḥ. al-āfiʿī (1181/1767) I, 289-304, *OMLT*, 132.

Ḥasan al-Gabartī, Badr al-Dīn Abu ʾl-Tadāʾī Ḥ. b. Ibrāhīm b. Ḥasan b. ʿAlī al-Zailāʾī al-G. alʿUqailī al-Ḥanafī (1188/1774) I, 385-408, *OALT*, 314/*OMLT*, 138.

Aḥmad al-Damanhūrī, ihāb al-Dīn Abu ʾl-ʿAbbās Aḥmad b. ʿAbd al-Munʿim b. Yūsuf al-Damanhūrī al-Azharī (1192/1778) II, 25, *OALT*, 317/*OMLT*, 142.

Aḥmad al-Suğāʾī, Abu ʾl-Faḍāʾil Aḥmad b. ihāb al-Dīn Aḥmad b. Muḥammad al-S. al-āfiʿī al-Azharī (1197/1783) II, 75, *OALT* 322/*OMLT*, 145.

al-Samannūdi, Muḥammad b. al-Hasan b- Muḥ. b. Aḥmad Gamāl al-Dīn b. Badr al-Dīn (1199/1785) II, 94, *OALT*, 326.

al-H̱aiyāṭ, Abu ʾl-Itqān Muṣṭafā al-Wafāʾī (1203/1789) II, 180, *OALT*, 372.

al-Ṣabbān, Abu ʾl-ʿIrfān Muḥammad b. ʿAlī al-Ṣ. al-Miṣrī al-Ḥanafī (1206/1791) II, 227, *OALT*, 379.

al-H̱ulaifī, Aḥmad b. Yūnus al-H̱. al-āfiʿī (1209/1794) II, 259-260, *OMLT*, 172.

al-Farġalī, āms al-Dīn Muḥammad b. ʿAbd Allah Fatḥ al-F. al-Sabarbāwī al-āfiʿī (1210/1795) II, 263, *OALT*, 388.

ÉDUCATION ET POLITIQUE AU XIXᵉ SIÈCLE : LES ÉLÈVES GRECS DANS LES GRANDES ÉCOLES D'INGÉNIEURS EN FRANCE

Fotini ASSIMACOPOULOU et Konstantinos CHATZIS

INTRODUCTION

En acquérant son indépendance en 1830, la Grèce accède dans les faits au statut d'un pays, certes libre, mais sans État organisé, en manque d'infrastructures élémentaires, marqué par une économie préindustrielle, avec comme seule tradition indigène en matière de connaissances techniques celle des corporations. Or, la modernisation du nouvel État, tant désirée par les gouverneurs du pays, nécessite des connaissances, des savoir-faire et une technologie qui dépassent de loin ces ressources autochtones. Devant ce vide, l'État grec s'applique très tôt à mettre en place un système de formation technique. En 1837, un jeune bavarois, l'officier du Génie Frédéric von Zentner, se voit chargé d'organiser à Athènes une école d'arts et métiers (l'ancêtre de l'actuelle École polytechnique d'Athènes, voir plus bas). Destiné initialement à former des ouvriers qualifiés et des contremaîtres dans le bâtiment, ce modeste établissement devient de proche en proche école de niveau supérieur qui décerne à ses élèves, à partir des années 1890, le diplôme d'ingénieur. Jusqu'à cette date, les seuls ingénieurs formés en Grèce sont les quelques officiers du Génie issus de l'École militaire que le premier gouverneur de Grèce, Jean Capodistrias, avait fondée en 1828. Mais la modernisation demande davantage. Des échanges tant " humains " que " matériels " de la Grèce avec l'Occident ne cessent alors de se multiplier tout au long du XIXᵉ siècle. Parmi ces échanges humains, ceux qui concernent les ingénieurs grecs ayant étudié à l'étranger occupent une place décisive. Si la présence de ceux-ci dans les divers chantiers de la modernisation (travaux publics, administrations techniques, enseignement technique, industrie...) est à plusieurs reprises mentionnée dans les travaux des spécialistes de l'histoire économique et technique de la Grèce, force est de constater qu'aucune " carte ", tant soit peu systématique, de cette population (origine sociale et géographique, carrière, diverses actions...) n'est actuellement disponible[1].

Cet article se veut une (toute première) touche à un tableau qui reste à dresser. Il s'appuie sur un travail de recensement des élèves grecs scolarisés dans les " grandes écoles " françaises d'ingénieurs au XIXᵉ siècle, établissements qui ont fourni à eux seuls une fraction importante des ingénieurs grecs ayant étudié à l'étranger à l'époque. Nous livrons ici les premiers résultats de cette enquête qui se poursuit[2].

NOMBRE ET ORIGINES DES ÉLÈVES GRECS

Durant le XIXᵉ siècle, plusieurs jeunes Grecs[3] prennent le chemin de Paris pour étudier l'art de l'ingénieur dans l'une (voire dans deux pour certains d'entre eux) des quatre écoles françaises d'ingénieurs les plus prestigieuses de l'époque : l'École polytechnique (EP), l'École des ponts et chaussées (EPC),

---

1. Sur le monde des ingénieurs grecs, voir pour le moment : Konstantinos Chatzis et Efthymios Nicolaïdis, " A pyrrhic victory : Greek women's conquest of a profession in crisis, 1923-1996 ", in Annie Canel, Ruth Oldenziel et Karin Zachmann (eds), *Crossing boundaries, building bridges. Comparing the history of women engineers, 1870s-1990s*, Amsterdam, Harwood Academic Publishers, 2000, 253-278 ; Konstantinos Chatzis, " Des ingénieurs militaires au service des civils : les officiers du Génie en Grèce au XIXᵉ siècle ", in Konstantinos Chatzis et Efthymios Nicolaïdis (eds), *Sciences, technologies et constitution de l'État au XIXᵉ siècle : le rôle des militaires*, Athènes, Centre de Recherches Néohelléniques (à paraître). On trouve dans ces deux articles des références sur les quelques travaux existant sur le sujet.

2. Pour notre travail de recensement, nous avons utilisé : pour l'École polytechnique la liste établie par Efthymios Nicolaïdis, " Les élèves grecs de l'École polytechnique (1820-1921) ", in Gilles Grivaud (éd.), *La Diaspora hellénique en France*, Athènes, École française d'Athènes, 2000, 55-65 ; pour les trois autres Écoles, les Registres Matricules des élèves (appropriés) de chaque établissement.

3. La constitution de l'échantillon " élève grec dans les Grandes Écoles d'ingénieur en France au XIXᵉ siècle" présente plusieurs difficultés, d'ordre pratique et conceptuel. Ainsi pour l'École des ponts et chaussées par exemple, nous ne disposons de registres d'inscription pour les élèves étrangers que pour la période postérieure à 1851 ; or, on sait, par ailleurs, que des Grecs étaient passés par cet établissement avant cette date. Mais à côté de ces problèmes pratiques, il y a la question plus délicate de la signification même du terme " élève grec " : celui-ci peut subsumer à l'époque sous un même vocable plusieurs populations : des élèves de nationalité grecque certes, mais aussi des sujets ottomans, des personnes originaires des îles ioniennes sous tutelle britannique, des Grecs nés à Londres, des élèves " valaques " etc. Pour les besoins de cet article, nous avons décidé de prendre au pied de la lettre la " volonté " des auteurs des *Registres* et de considérer comme Grecs tous ceux qui sont répertoriés comme tels (sous le vocable de Grec ou d'Hellène) par les instances des quatre établissements traités ici, tout en incluant quelques sujets " valaques " et " moldaves " (de l'ordre d'une demi-douzaine) qui semblent appartenir à des familles grecques (tels que les Doukas ou les Rosetti, par exemple : voir Neagu Djuvara, *Le pays roumain entre Orient et Occident*, Paris, Publications orientalistes de France, 1989, 36). Des analyses statistiques pour les élèves dont on connaît les origines géographiques (ce qui n'est pas le cas pour l'ensemble des élèves répertoriés) feront apparaître le poids de chaque population.

Vu les doutes qui subsistent sur le caractère exhaustif de l'échantillon construit ainsi que sur l'inclusion ou pas de certains élèves, les résultats ici présentés sont susceptibles de corrections et d'affinements postérieurs. Cela étant, nous pouvons d'ores et déjà affirmer que les grandes tendances dégagées sont solidement établies.

l'École des mines (EM) et l'École centrale des arts et manufactures (ECAM)[4].
En effet, pour la période 1830-1912[5], nous avons pu répertorier 193 inscriptions d'élèves grecs[6], correspondant à 177 inscrits, étant donné que 16 d'entre
eux optent pour un double cursus, à l'École polytechnique d'abord, dans l'une
de ses écoles d'application (EPC ou EM) ensuite.

C'est l'École des ponts[7] qui reccueille le suffrage du plus grand nombre
avec 67 Grecs inscrits pendant la période 1851-1912 (dont 6 étant passés par
l'École polytechnique)[8]. Elle est talonnée par l'École polytechnique, plébisci-
tée quant à elle par 55 Grecs[9]. L'École des mines affiche, pour la période 1830-

---

4. Pour une vue générale sur le système de formation des ingénieurs en France au XIXe siècle,
voir Antoine Picon et Konstantinos Chatzis, " La formation des ingénieurs français au siècle dernier. Débats, polémiques et conflits ", *L'orientation scolaire et professionnelle*, vol. 21, n° 3 (septembre 1992), 227-243 ; Ulrich Pfammatter, *The making of the modern architect and engineer*,
Bâle, Birkhäuser, 2000 ; Robert Fox et George Weisz (eds), *The organization of science and technology in France, 1808-1914*, Cambridge, Cambridge University Press ; Paris, Editions de la Maison des Sciences de l'Homme, 1980. Sur les spécificités du système français de formation des
ingénieurs par rapport à d'autres systèmes nationaux, voir Peter Lundgreen, " Engineering education in Europe and the USA, 1750-1930 : the rise to dominance of school culture and the engineering professions ", *Annals of Science*, 47 (1990), 33-75.

5. Deux mots sur les bornes extrêmes de la période retenue ici (1830-1912). 1830 est une date
" naturelle ", elle correspond à la fondation de l'État grec. 1912 inaugure une période de guerres
(1912-1922) à la fin desquelles la Grèce voit son territoire et sa population doubler et une grande
partie de l'hellénisme vivant hors des frontières de l'État grec s'installer définitivement à l'intérieur des frontières nationales. Certes, durant la période 1830-1912, le territoire national s'est
élargi à deux reprises, en 1864 quand l'Angleterre cède à l'État grec les îles ioniennes, et en 1881,
suite à l'annexion de la Thessalie et d'une partie de l'Épire. Mais ces élargissements n'ont rien de
comparable avec les bouleversements qui ont suivi la décennie 1912-1922.

6. Pour établir quelques points de comparaison, à l'École polytechnique de Zurich (créée en
1855), autre lieu de destination privilégiée pour des Grecs désirant devenir ingénieurs, on trouve
jusqu'en 1912, une quarantaine de Grecs inscrits en Génie civil et une dizaine en Génie mécanique
et Génie électrique (estimations d'après l'*Epetiris ton foitisanton Ellinon eis to Omospondiakon
Polytechneion tis Zyrichis apo tis idryseos tou mechri 31/12/1981 (Annuaire des élèves grecs
ayant étudié à l'École polytechnique de Zurich, de sa fondation au 31 déc. 1981)*, s.d.). Kostas
Lappas dans sa thèse de doctorat sur l'Université d'Athènes au XIXe siècle (*Panepistimio kai foitites stin Ellada kata ton 19o aiona*, thèse de doctorat de l'Université d'Athènes, 1997, 394-395)
recense également quelques travaux consacrés aux Grecs qui ont étudié à l'étranger durant le XIXe
siècle. Sur les étudiants grecs en Allemagne dans toutes les disciplines sauf celles de l'ingénieur
et de l'architecte, voir Konstantina Zorbala, " Ellines foitites sta germanika panepistimia kata ton
19o aiona ", *Les temps de l'histoire. En vue d'une histoire de l'enfance et de la jeunesse* : actes
du colloque international, Athènes, 17-19 avril 1997, Athènes, Archives historiques de la jeunesse
grecque (Secrétariat Général de la Jeunesse), 1998, 55-62.

7. Sur les conditions d'admission, le statut et la scolarité des étrangers à l'École des ponts et
chaussées, voir Konstantinos Chatzis, " Die älteste Bauingenieurschule der Welt - die École des
ponts et chaussées (1747-1997) ", *Bautechnik Spezial*, 1998, 26-42.

8. Cette liste établie par nos soins doit être complétée par 6 autres Grecs admis à l'École des
ponts entre 1825 et 1850 et dont l'identité nous est inconnue. Voir Emile Malézieux, " Les élèves-
externes de l'École des ponts et chaussées ", *Annales des Ponts et Chaussées*, t. IX (1875), 1er
semestre, 5-23.

9. Voir E. Nicolaïdis, " Les élèves grecs de l'École polytechnique (1820-1921) ", *op. cit.* E.
Nicolaïdis a établi une liste de 57 élèves grecs admis à l'EP pour la période 1794-1921 (dont 53
pour la période 1830-1912). Dans une excursion aux archives de l'École polytechnique, nous sommes tombés, par hasard, sur deux autres élèves grecs avec la mention *rayé* [des cadres de l'École] :
Aristidi Delyanni (1896) et Markades (1841-1842).

1912, un score honorable avec 46 élèves inscrits (parmi lesquels 10 avaient entrepris, au préalable, des études à l'École polytechnique). Elle laisse ainsi loin derrière l'École centrale, qui compte parmi ses diplômés, pour la même période, seulement 25 Grecs[10]. Signalons qu'une inscription ne se traduit pas automatiquement par un diplôme, plusieurs Grecs admis ayant dû quitter ces établissements sans aucune récompense institutionnelle. C'est le cas, par exemple, d'une quinzaine de Grecs qui, après avoir tenté leur chance à l'École des ponts, se voient sommés de la quitter pour cause d'insuffisance de travail, ou démissionnent pour diverses raisons (dont la maladie).

Une simple mise sous forme graphique du nombre des élèves admis aux quatre écoles (figures 1 à 5) met en évidence, pour l'ensemble de la période 1830-1912, l'existence d'un mouvement d'inscriptions à flux quasi continu — il n'y a pratiquement pas d'année sans inscription —, marqué par trois " vagues " d'ampleur inégale qui se situent aux alentours des années 1840, 1860 et 1890 respectivement.

La première arrivée en nombre d'élèves grecs se produit au début de la période et elle est entièrement polytechnicienne : entre 1837 et 1842, quatorze Grecs sont admis à l'École polytechnique dont un continuera ses études à l'École des mines[11]. La deuxième " vague " d'entrées coïncide avec la fin du règne du roi Othon et le changement de dynastie : d'ampleur comparable à la précédente, puisque entre 1859 et 1863 on enregistre 19 inscriptions (pour 15 inscrits) dans les quatre écoles, cette deuxième vague d'admissions est toujours dominée par le poids de l'École polytechnique qui attire, à elle seule, 11 Grecs. La troisième série d'arrivées, de loin la plus importante pour notre période, est marquée par un déplacement dans le choix des élèves : l'École polytechnique perd, de façon nette, du terrain au profit des trois autres établissements, qui dispensent un enseignement plus appliqué. Durant la décennie 1885-1895, le nombre des Grecs admis dans les quatre établissements s'élève en effet à 71 (pour 76 inscriptions). Trente-six d'entre eux (dont 3 anciens polytechniciens) optent pour l'École des ponts, 20 (dont 2 polytechniciens) choisissent l'École des mines, alors que l'École centrale et l'École polytechnique, occupant ex-aequo la dernière place dans les choix des candidats, recueillent chacune le suffrage de 10 Grecs.

---

10. Pour l'École centrale des arts et manufactures, nous n'avons pu répertorier pour le moment que les *élèves diplômés*.

11. Sur les conditions d'admission, le statut et la scolarité des étrangers à l'École polytechnique durant le XIXe siècle, voir Anousheh Karvar, *La formation des élites scientifiques et techniques étrangères à l'École polytechnique aux XIXe et XXe siècles*, Thèse de doctorat nouveau régime, Université de Paris VII, 1997. Sur l'enseignement à l'École des mines (admission des élèves, cours préparatoires, examens, diplômes et certificats d'études, voir Adolphe Carnot, " Notice sur l'enseignement de l'École Nationale Supérieure des mines ", *Annales des Mines, Mémoires*, t. XV, 8e série (1889), 70-89.

Revenons sur ces vagues d'admissions et essayons d'en analyser les principaux ressorts. La première semble être directement liée aux efforts du tout jeune royaume hellénique pour se transformer en État moderne à l'occidentale. L'un des moyens mis au service de cette cause par les gouvernants de l'époque est l'envoi de boursiers à l'étranger. En effet, parmi les admis à l'École polytechnique de cette période figurent plusieurs anciens élèves de l'École militaire grecque (Scholi Evelpidon), tels que Emmanouil Isaïas (Manuel d'Isay) (EP 1842) (que l'on retrouve à l'École des mines), Efstathios Tomaropoulos (EP 1837), Dimitrios Koromilas (EP 1837) et Dimitrios Skalistiris (EP 1838), grande figure du monde des ingénieurs grecs au XIXe siècle (voir plus bas). Ayant déjà acquis les premières connaissances dans le domaine du génie civil lors de leur passage à l'École militaire, ces jeunes Grecs se voient envoyés par l'État grec en France afin de parfaire leurs études et assurer dans les meilleures conditions leurs fonctions dans les domaines de l'aménagement des villes, des travaux publics et de la construction de bâtiments publics[12].

La deuxième période d'arrivée d'élèves grecs semble également procéder de la même volonté de modernisation exprimée par l'État grec, qui expérimente, dans les années 1850, plusieurs mesures visant le développement du pays. À partir des années 1850, Athènes tente de mettre en place son plan d'aménagement, d'autres villes commencent timidement à s'équiper en bâtiments publics et en infrastructures élémentaires. En 1852, une importante loi relative aux routes voit le jour, et l'État se lance dans la construction d'un premier réseau (qui reste très insuffisant)[13]. En 1855, la décision de construire la liaison ferrée Athènes-Pirée est votée par le parlement grec. Deux ans plus tard, l'ingénieur des ponts et chaussées Paul-Ernest Daniel est invité par le gouvernement Boulgaris, avec comme mission d'organiser le Service des travaux publics au sein du Ministère de l'intérieur. En 1858, une commission se penche sur le problème de manque de techniciens-géomètres en Grèce. Cet intérêt pour les travaux publics survit au règne d'Othon, puisque le gouvernement provisoire qui s'installe juste après la destitution du roi, survenue en 1862, décide d'élever le niveau de l'enseignement technique dispensé jusqu'alors à l'École des arts et métiers d'Athènes (l'actuelle École polytechnique). On comprend aisément qu'un tel contexte est favorable pour des études d'ingénieur à l'étranger. L'État grec montre le chemin en envoyant en France plusieurs de ses sujets. Parmi les 12 Grecs qui se sont assis durant les années 1859-1864 sur les bancs de l'École polytechnique figurent, en effet, 7 officiers du Génie[14] et un autre boursier

12. Rappelons que jusqu'en 1878, les officiers du Génie remplissent, en Grèce, dans les faits des fonctions et des tâches qui relèvent de l'ingénieur des ponts et chaussées. Voir K. Chatzis, " Des ingénieurs militaires au service... ", *op. cit.*

13. Sur ce sujet voir Maria Synarelli, *Dromoi kai limania stin Ellada, 1830-1880* (*Des routes et des ports en Grèce, 1830-1880*), Athènes, ETVA, 1989.

14. Il s'agit de : Nikolaos Nikolaïdis (EP 1861), Ioannis Sechos (EP 1862), Nikolaos Metaxas (EP 1862), Alexandros Fountouklis (EP 1862), Dimitrios Karageorgis (EP 1862), Ioannis Markopoulos (EP 1863), Nikolaos Tsamados (EP 1864).

d'État, le diplômé de l'université d'Athènes Anastasios Soulis (EP 1861).
Notons enfin la présence, parmi les élèves de cette deuxième vague, d'Antoine
Matsas (EP 1859) qui va s'illustrer comme grand entrepreneur dans le secteur
des travaux publics en Grèce.

La dernière vague d'inscriptions enfin coïncide avec l'ère du ministère de
Charilaos Trikoupis en Grèce (1882-1895), qui est également celle du " dernier
sursaut " de l'Empire ottoman[15]. Époque des grands travaux pour les deux
régions, comme on peut en juger d'après les chiffres suivants.

Alors que durant l'époque d'Othon (1833-1862), l'État grec consacre aux
travaux publics à peine 7,3 millions de drachmes (l'ensemble des dépenses de
l'État s'élevaient pour cette période à 470 millions), les seules années 1879-
1890 voient le développement de 2200 km de réseau routier, qui absorbe quelque 65 millions de drachmes. Le canal de Corinthe (1882-1893) coûte au total
60 millions de drachmes[16], et la construction, entre 1883 et 1892, de près de
900 km de réseaux de chemins de fer est d'un coût égal à 144 millions de drachmes[17]. Marquée par les grands travaux du ministère de Trikoupis, la Grèce
du dernier quart du XIXe siècle connaît également le développement du secteur
minier, les mines de Laurium en étant une figure emblématique[18].

Quant à l'Empire ottoman, il devient, durant le dernier quart du XIXe siècle,
l'objet d'un véritable engouement de la part des capitaux étrangers (les capitaux français arrivant en tête des investissements étrangers : s'élevant à 85 millions de francs en 1881, ils passent à 292 millions en 1895 et à 511 millions
en 1909). Ces années correspondent, entre autres, à une construction accélérée
des chemins de fer dans l'Empire, domaine de prédilection des investissements
étrangers. Alors que l'on comptait 1.800 kilomètres de voies ferrées en 1878,
il y en aura 5.800 en 1908. On sait, par ailleurs, que ce phénomène d'expansion économique de l'Occident dans l'Empire s'accompagne de l'ascension
des minorités non musulmanes, surtout des Grecs et des Arméniens, dont la
position est désormais dominante non seulement dans le commerce extérieur
mais aussi dans les activités bancaires et industrielles (le secteur minier y com-

15. Pour utiliser l'expression de François Georgeon, " Le dernier sursaut (1878-1908) ", in
Robert Mantran (dir.), *Histoire de l'Empire Ottoman*, Paris, Fayard, 1989, 523-576.

16. Sur le percement du canal et ses retombés économiques, voir Evi Papayannopoulou, *I Dio-
ryga tis Korinthou* (*Le canal de Corinthe*), Athènes, ETVA, 1989.

17. Pour ces données (et des références bibliographiques), voir : D. Dakin, *The Unification of
Greece, 1770-1923*, London, Ernest Benn Limited, 1972 ; K. Chatzis, " Des ingénieurs militaires
au service des civils... ", *op. cit.* L'ouvrage de référence sur les travaux publics à l'époque de
Trikoupis est désormais Lidya Tricha, *Charilaos Trikoupis et les travaux publics* (éd. bilingue,
grec-français), Athènes, Kapon, 2001. Sur les chemins de fer l'ouvrage spécialisé de Lefteris
Papayiannakis, *Oi ellinikoi sidirodromoi, 1882-1910* (*Les chemins de fer grecs, 1882-1910*), MIET,
1982.

18. Signalons la présence, plus qu'importante, dans tous ces secteurs de capitaux étrangers
(dont ceux des Grecs de l'Empire ottoman et de la Diaspora), capitaux qui montrent une préfé-
rence marquée pour les investissements indirects, surtout par le biais de la dette publique.

pris). En effet, le premier recensement industriel de l'Empire ottoman réalisé en 1913-1915 montre que 50% du capital investi dans les établissements industriels recensés sont aux mains des Grecs[19].

Dans un tel contexte, il n'est pas étonnant de voir de nombreux Grecs, de nationalité grecque ou faisant partie de l'hellénisme de l'Empire ottoman, venir massivement gonfler les rangs des écoles d'ingénieurs en France. Comme par le passé, l'État grec, à l'origine des grands travaux publics de la période, renvoie en France des fonctionnaires[20], pour l'essentiel issus de l'École polytechnique d'Athènes qui est devenue école de niveau universitaire en 1887. Mais c'est la " société civile ", désormais très active, qui est l'acteur le plus dynamique dans ce mouvement d'immigration " intellectuelle " vers les écoles françaises. Nombreux sont alors les jeunes Grecs qui se rendent, de leur propre chef, en France, pour étudier l'art de l'ingénieur.

Comment expliquer, après la recrudescence des années 1885-1895, la diminution plus que substantielle du nombre des élèves grecs ayant étudié dans les quatre écoles françaises ici traitées après 1900 ? Plusieurs causes conjuguent leurs actions pour produire cet effet, pendant une période marquée, entre autres, par la faillite de l'État grec en 1893, sa défaite lors du conflit gréco-turc en 1897 et l'imposition d'une Commission internationale de contrôle des finances du pays en 1898. Contentons-nous d'évoquer celles qui sont liées à l'offre d'études d'ingénieur. À partir de la fin du XIXᵉ siècle, les quatre établissements étudiés ici doivent affronter une triple concurrence. Concurrence " française " d'abord, avec la création à la fin du XIXᵉ siècle et au début du XXᵉ siècle d'une nouvelle génération d'établissements d'ingénieurs, spécialisés dans des disciplines liées à la deuxième révolution industrielle (électricité, chimie etc.). Concurrence " allemande " ensuite, les écoles techniques de langue allemande (Suisse alémanique comprise) faisant preuve à partir des années 1870 d'un dynamisme et d'une excellence mondialement reconnus. Concurrence " grecque " enfin : devenue établissement de niveau supérieur à partir de 1887, en grande partie grâce à la contribution des ingénieurs grecs issus de nos quatre écoles, l'École polytechnique d'Athènes offre désormais la possibilité aux jeunes Grecs attirés par les études techniques de devenir ingénieurs sans avoir à se déplacer à l'étranger : de 1890 à 1912, les deux départements en exercice de l'École polytechnique d'Athènes délivrent 261 diplômes en Génie civil et 29 diplômes en Génie mécanique[21].

19. Pour ces chiffres, voir F. Georgeon, " Le dernier sursaut (1878-1908) ", *op. cit.*

20. L'École des ponts, destination privilégiée pour les boursiers de l'État grec, reçoit ainsi, pendant la période 1881-1912, 19 fonctionnaires.

21. Calculs d'après le tableau contenu dans Technikon Epimelitirion tis Ellados (Chambre technique de Grèce), *Techniki Epetiris tis Ellados* (*Annuaire technique de la Grèce*), t. A, vol. I, Athènes, Ekdoseis tou Technikou Epimelitiriou tis Ellados, 1935, 62.

Dernière question enfin de cette partie. D'où viennent ces jeunes grecs désirant d'étudier l'art de l'ingénieur en France ? Quelles sont leurs origines géographiques et sociales ?

Sur un échantillon de 141 élèves grecs, pour lesquels on a pu identifier le lieu de naissance, les sujets grecs et les Grecs des divers " hellénismes périphériques " se trouvent à égalité (71 contre 70). Une analyse plus fine montre que parmi les 70 Grecs nés en dehors des frontières de l'État grec de l'époque, 5 sont originaires des îles ioniennes avant que celles-ci ne deviennent grecques en 1864 et 5 autres sont nés en Thessalie avant son annexion à l'État grec en 1881. Parmi les 60 élèves restant, 14 sont issus de familles installées dans des pays situés autour de la Mer Noire (Bulgarie, Roumanie et Russie), 5 sont nés dans des villes de l'Europe occidentale, 3 appartiennent à l'hellénisme d'Égypte, 1 est né à Beyrouth. Trente-sept enfin sont des fils de familles vivant à l'intérieur des frontières de l'Empire ottoman (13 sont nés à Istanbul).

L'enquête sur les origines sociales des élèves n'a pu aboutir que pour 49 élèves (dont 29 nés en Grèce), pour lesquels on a identifié la profession du père. Même si la taille réduite de notre échantillon enlève aux conclusions que l'on peut en tirer toute représentativité statistique, tentons quelques remarques, largement ouvertes à la révision à la lumière de nouvelles données. Premier fait saillant à signaler, le poids de la bourgeoisie économique : plus de la moitié des élèves sont issus de familles qui appartiennent à ce milieu social[22]. Signalons également, pour le cas de Grecs nés en Grèce, que si la moitié d'entre eux (13 sur 29) sont issus de la bourgeoisie économique, les professions libérales (des médecins (3), des avocats (1), ou des journalistes (2), par exemple) ainsi que celles liées à l'appareil d'État (des militaires (3), un ministre, un procureur du roi, un professeur de littérature ancienne, un employé au Ministère de la marine...) ne dédaignent pas d'envoyer leur progéniture à l'étranger pour étudier le métier d'ingénieur[23]. Notons enfin que le système de bourses mis en place par l'État grec a permis au fils d'un modeste tailleur de Volos (il s'agit

22. Pour construire la catégorie " bourgeoisie économique ", nous avons regroupé les professions suivantes : négociant (11 occurrences), rentier (5), commerçant (4), propriétaire (3), marchand (1), industriel (1), armateur (1), banquier (1). Nous n'ignorons pas le débat historiographique, dans lequel sont impliqués en France des historiens comme Terry Shinn, William Serman, Christophe Charle et Bruno Belhoste, sur la signification économique et sociale de termes comme " commerçant ", " propriétaire " ou " rentier " (on sait par exemple que le statut de rentier ou de propriétaire est revendiqué au XIXe siècle en France par des personnes de rang et de fortune modestes). Cela étant, le cas grec nous paraît plus facile à traiter. Tout d'abord, les informations dont on dispose par ailleurs sur un certain nombre de familles dont les élèves traités ici sont issus montrent qu'il s'agit bien des familles " bourgeoises ". Qui plus est, le fait même de pouvoir envoyer sa progéniture pour étudier à l'étranger plaide en faveur de l'hypothèse selon laquelle il s'agit bien de familles riches. Nous aimerions remercier à cette occasion B. Belhoste qui a aimablement mis à notre disposition le manuscrit de son livre (à paraître prochainement) sur l'École polytechnique et ses élèves, de la Révolution au Second Empire.

23. Pour la période 1900-1940, voir l'étude de Aliki Vaxevanoglou, *Oi Ellines kefalaiouchoi 1900-1940 : koinoniki kai oikonomiki prossegkissi* (*Les capitalistes grecs 1900-1940 : une approche économique et sociale*), Athènes, Themelio, 1994.

de Ioannis Raptakis : EPC 1885) d'accéder au diplôme de l'École des ponts et chaussées avant d'entamer une longue carrière de professeur à l'École polytechnique d'Athènes (sur Raptakis, voir plus bas).

ACTIONS ET CARRIÈRES

Les élèves grecs et la diffusion des connaissances techniques

Un premier domaine où les ingénieurs grecs passés par les grandes écoles d'ingénieurs en France au XIXᵉ siècle vont s'illustrer est celui de la transmission en Grèce des connaissances et savoir-faire élaborés en Occident. Plusieurs d'entre eux, une fois rentrés en Grèce, s'installent à la tête des établissements dispensant, à l'époque, un enseignement technique, assurent une grande partie de l'enseignement et publient plusieurs traités techniques[24].

Seule école d'ingénieurs du royaume hellénique jusqu'en 1887, l'École militaire " embauche " régulièrement pour ses besoins d'enseignement, tout au long du XIXᵉ siècle, plusieurs de ces Grecs passés par la France[25]. Ainsi l'officier du Génie Dimitrios Skalistiris (EP 1838) y enseigne, dans les années 1850, les routes et les ponts. Ses confrères Alexandros Fountouklis (EP 1862), Ioannis Sechos (EP 1862) et Nikolaos Nikolaïdis (EP 1861 et EPC 1862) professent, dans les années 1860, les machines, les travaux publics et les mathématiques, respectivement. L'officier de l'artillerie Vlassis Valtinos (EP 1850) enseigne, toujours dans les années 1860, la géographie et l'histoire militaires. Mais l'École ne se contente pas de recruter des militaires. Des " civils ", passés par la France, comme le polytechnicien Théodoros Negris (il sera professeur de mathématiques) et Dimitrios Stroumbos (EP 1831), professeur de physique et de mécanique, figurent pendant longtemps parmi les membres de son corps enseignant (notons ici que Stroumbos a marqué également la jeune Université d'Athènes puisque il y enseigne de 1844 à 1890).

Le rôle des Grecs ayant étudié dans les grandes écoles françaises d'ingénieurs n'est pas moindre au sein de l'autre établissement dispensant une formation technique en Grèce durant le XIXᵉ siècle, l'actuelle École polytechnique d'Athènes (en grec : *Ethniko Metsovio Polytechneio*)[26].

---

24. Grâce à un projet de l'École polytechnique d'Athènes, auquel Fotini Assimacopoulou collabore, les résultats d'un recensement de ces traités seront bientôt disponibles.

25. Voir K. Chatzis, " Des ingénieurs militaires au service des civils... ", *op. cit.*

26. Les informations qui suivent sur cet établissement et ses enseignants sont tirées de Konstantinos Biris, *Istoria tou Ethnikou Metsoviou Polytechneiou* (*Histoire de l'École polytechnique d'Athènes*), Athènes, 1957 ; voir aussi Angheliki Fenerli, " Spoudes kai spoudastes sto Potytechneio, 1860-1870 " (" Études et étudiants de l'École Polytechnique, 1860-1870 "), *Université : Idéologie et culture, dimensions historiques et perspectives* : Actes du Colloque international, Athènes, 21-25 septembre 1987, tome I, Athènes, Archives historiques de la jeunesse grecque, Secrétariat général à la jeunesse, 1989, 151-166 et Hélène Kalaphati, " To Ethniko Metsovio Polytechneio sto gyrisma tou aiona : epaggelmatikes dieksodoi ton apofoiton kai thesmiko kathestos tou idrymatos " (" L'École Polytechnique Nationale au tournant du siècle : débouchés professionnels des diplômés et statut institutionnel de l'établissement "), *ibid.*, 167-183.

Celle-ci démarre en 1837 comme modeste école d'arts et métiers destinée à former, grâce à des cours du dimanche, des ouvriers qualifiés et des contremaîtres dans le bâtiment. À sa tête, de 1844 à 1862, l'architecte L. Kaftantzoglou renforce l'orientation artistique de l'établissement aux dépens de sa vocation technique initiale. Mais pas pour longtemps. Pour répondre au nouveau contexte des années 1850 (voir *supra*), l'établissement change à la fois de statut et d'orientation. Le rôle des ingénieurs grecs passés par la France sera déterminant dans ce processus d'élévation continu du niveau d'enseignement technique dispensé au sein de cette école, qui devient à partir de 1887 un établissement de niveau universitaire. Ioannis Papadakis (EP 1842 et EM 1844), professeur d'astronomie à l'Université d'Athènes, codirecteur (avec Stamatios Krinos et Gerasimos Metaxas) de l'établissement durant les années 1863-1864, donne les premières impulsions en faveur de l'élévation du niveau d'enseignement. L'officier du Génie, Dimitrios Skalistiris (EP 1838), directeur de l'école de 1864 à 1873, relaye son action. Son successeur, l'officier du Génie Dimitrios Antonopoulos (1873-1876) est aussi un ancien de l'École polytechnique de Paris (EP 1848). À l'instar de l'École militaire, l'institution fait, elle aussi, massivement appel aux Grecs formés dans les grandes écoles d'ingénieurs en France pour constituer son corps enseignant. Outre ses fonctions de direction déjà mentionnées, I. Papadakis assure pendant longtemps l'enseignement de la Géométrie descriptive (1853-1856 et 1863-1876). L'officier du Génie Leonidas Vlassis (EP 1850) participe à l'élévation du niveau des études en professant les principes de statique et de mécanique pendant une période brève mais cruciale dans l'histoire de l'établissement (1862-1864). Dans les décennies suivantes, on trouve, parmi les enseignants de l'établissement, deux autres officiers du Génie : D. Skalistiris comme professeur de mécanique d'abord, des routes et des ponts ensuite (1864-1883), et Ioannis Sechos (EP 1862) qui enseigne d'abord la statique, la mécanique et l'hydraulique puis la construction (1869-1888). Mais assez vite les civils vont prendre la succession des militaires dans les fonctions d'enseignement. Ainsi, Anastasios Soulis (EP 1861 et EPC 1863) enseigne pendant une longue période (1874-1910) la mécanique d'abord, la résistance des matériaux et l'hydraulique ensuite : il sera, par ailleurs, le premier à enseigner en Grèce la théorie du béton armé. Antonios Damaskinos (EP 1859) professe la trigonométrie et l'analyse (1877-1884). Diplômé de l'École centrale des arts et manufactures, Iason Zochios (ECAM 1866) sera pendant trois ans professeur de mécanique appliquée et de chemins de fer (1881-1884), avant de céder sa place à Nikolaos Gazis (EPC 1879) (1884-1888). Ioannis Argyropoulos (EM 1872) marque de sa présence l'établissement en professant, de 1889 à 1908, les chemins de fer et, de 1908 à 1923, la métallurgie. Il en va de même de Ioannis Raptakis (EPC 1885) qui enseigne, de 1890 à 1917 et de 1920 à 1923, la construction et la statique graphique ainsi que de Georgios Maltezos, ancien élève de l'École des mines (EM 1893), qui familiarise les élèves, de 1896 à 1938 (avec quelques interruptions), avec le fonctionnement des machines (dont la machine à vapeur). Dimosthenis Protopapadakis (EPC 1894) fera aussi une belle carrière au sein de l'établissement, comme professeur des chemins de fer de 1908 à 1917 et de 1920 à 1948, tout en assurant également de 1933 à 1935 la direction de l'école. D'autres diplômés de nos quatre écoles françaises vont apporter leur contribution à la marche de l'École polytechnique. Ilias Aggelopoulos (EPC 1883), grande figure du monde des ingénieurs grecs (voir plus bas), professe la construction de 1888 à 1890 et les ponts et routes en 1897 et 1898. Petros Protopapadakis (EP 1881 et EM 1883) est professeur de travaux portuaires et hydrauliques de 1889 à 1891. Signalons enfin les brefs passages de Ioannis Markopoulos (EP 1863 et EPC 1864), qui enseigne en 1888 la construction, et de Nikolaos Sideridis (EPC 1880) qui donne des cours de façon éphémère les travaux portuaires en 1889.

Les élèves grecs au service de l'Administration[27]

De 1829 jusqu'aux années 1870, le corps du Génie, alimenté par des officiers sortis de l'École militaire, remplit dans les faits des fonctions et des tâches qui relèvent des domaines de l'ingénieur des ponts et chaussées. Les officiers du Génie envoyés par les gouvernements successifs en France pendant cette période, une fois rentrés au pays, vont jouer alors un rôle très actif comme ingénieurs d'État dans les domaines des travaux publics, de l'urbanisme et de la construction des bâtiments publics[28]. En 1878, le gouvernement d'Alexandros Koumoundouros vote la création d'un corps d'ingénieurs des ponts et chaussées dépendant du Service des travaux publics au sein du Ministère de l'intérieur. Si le nouveau corps est initialement alimenté pour l'essentiel d'ex-officiers du Génie, il va vite attirer plusieurs jeunes Grecs ayant étudié l'art de l'ingénieur à l'étranger, en particulier la France.

Le premier directeur du nouveau corps est un ancien de l'École polytechnique, à maintes reprises mentionné dans ce texte, D. Skalistiris (1878-1883). Son successeur est un autre ancien polytechnicien (et diplômé de l'École des ponts), A. Soulis (1883-1885), tout comme le directeur suivant, l'officier du Génie L. Vlassis (1885-1886). Dix ans après la création du nouveau corps, en 1888, les ingénieurs portant le titre d'ancien élève d'une des quatre écoles françaises ici traitées représentent le *tiers des effectifs*[29]. Dans les années qui

27. Sauf mention explicite, les informations sur les carrières des ingénieurs grecs passés par l'EPC, l'EM et l'ECAM sont tirées des annuaires publiés par l'*Association des anciens élèves* de chaque établissement (différentes années) ainsi que de : Technikon Epimelitirion tis Ellados (Chambre technique de Grèce), *Techniki Epetiris tis Ellados* (*Annuaire technique de la Grèce*), t. B, Athènes, Ekdoseis tou Technikou Epimelitiriou tis Ellados, 1934.

28. Pour un bilan de l'action des officiers du Génie, voir K. Chatzis, " Des ingénieurs militaires au service des civils... ", *op. cit.*

29. Il s'agit de : Leonidas Vlassis (EP 1850), Anastasios Soulis (EP 1861 et EPC 1863), Ioannis Markopoulos (EP 1863 et EPC 1864), Dimitrios Soutsos (ECAM 1872), Pavlos-Sokratis Omiros (EPC 1880), Nikolaos Balanos (EPC 1880), Nikolaos Gazis (EPC 1879), Ioannis Argyropoulos (EM 1872), Ilias Aggelopoulos (EPC 1883), Dimitrios Aravantinos (EPC 1883), Nikolaos Sideridis (EPC 1880), Athanasios Georgiadis (EPC 1882).Voir *Michaniki Epitheorisi*, n° 7-8 (janvier-avril 1888), 48-49.

suivent, d'autres Grecs étant passés par la France vont mettre leurs compétences, de façon définitive ou provisoire, au service de l'État grec. C'est le cas de : Andreas Igglessis (EPC 1884), Ioannis Raptakis (EPC 1885), Konstantinos-Petros Xydis (EPC 1886), Dimitrios Tzogias (EPC 1890), Alexandros Oikonomos (EPC 1894), Aristidis Balanos (ECAM 1889), Athanasios-Ilias Kantas (ECAM 1888). Konstantinos Argyropoulos (ECAM 1908), après une brève carrière à la Compagnie hellénique d'électricité (1911-1914), se trouve à partir de 1915 au Département Électrique du Ministère des transports. Pavlos Kampas (EPC 1910) suit le chemin inverse : ingénieur d'État pour la période 1915-1922, il exerce en libéral à partir de 1922.

Notons aussi qu'un nombre (limité) de nos ingénieurs travaillent — durant une partie de leur carrière du moins — pour le compte de municipalités. C'est le cas de Pieris Kalichiopoulos (ECAM 1851), qui sera pendant longtemps ingénieur municipal (et même directeur) au service des eaux de son île natale, Corfou, avant de s'installer, en fin de carrière, à son propre compte. C'est le cas aussi de P. Protopapadakis et I. Aggelopoulos, qui effectuent de brefs passages aux services techniques de la ville d'Athènes. Panagos Kaperonis (EPC 1891) travaille également, de 1895 à 1899, pour la ville d'Athènes avant de fonder son propre bureau d'études. Georgios Menexès, originaire de Salonique, admis à l'École des ponts et chaussées en qualité de fonctionnaire grec en 1885 après avoir fait des études à l'École polytechnique d'Athènes, sera pendant longtemps ingénieur en chef de sa ville natale avant d'exercer en libéral.

Les élèves grecs dans le secteur privé

Si plusieurs des ingénieurs grecs formés dans les grandes écoles françaises d'ingénieur investissent les départements techniques de l'État grec et le système éducatif du pays, nombreux sont aussi ceux qui exercent leurs compétences dans le secteur privé (c'est le cas notamment des ingénieurs grecs sortis de l'École centrale et de l'École des mines). Notons aussi que l'aller-retour entre l'Administration et le secteur privé semble être un phénomène de plus en plus fréquent à partir des années 1900, ce qui rend la distinction ingénieur d'État / ingénieur " civil " difficile à manier.

Le secteur minier, en Grèce comme dans l'Empire ottoman, est un débouché naturel pour les diplômés de l'École des mines mais aussi pour un certain nombre de Grecs ayant étudié à l'École centrale. Parmi les nombreuses mines de la région, celles de Laurium sont un passage " obligé " pour plusieurs des ingénieurs passés par la France, tels que Stéphanos Xydias (EM 1887), Georgios et Achilleas Georgiadis (EM 1885 et 1887), Ioannis Argyropoulos (EM 1872). Certains n'ont pas peur de se déplacer loin de leur lieu d'origine : Tsapalos (EM 1895) travaillera dans les mines espagnoles et A. Georgiadis dans celles de la Sardaigne.

Les compagnies de chemins de fer attirent aussi plusieurs diplômés. Pana-giotis Rodios, fils de colonel, et Michalis-Ioannis Agelastos, fils d'agent de change installé à Londres, diplômés de l'École centrale en 1887 et 1888 res-pectivement, démarrent leur carrière comme ingénieurs à la *Société internatio-nale anonyme de construction et d'entreprises de travaux publics* (il s'agit, selon toute vraisemblance, de la *Diethnis Etaireia Oikodomon kai Ergolavion Dimosion Ergon*, compagnie qui assure la construction et l'exploitation de plu-sieurs réseaux de chemin de fer en Grèce durant cette période). Agelastos fera une carrière cosmopolite dans les chemins de fer (voir plus bas), quant à Rodios, il est au début des années 1900 sous-chef de la traction au chemin de fer Pirée-Athènes-Peloponnèse. Le syriote Georgios Doumas (EPC 1890), fils de Ioannis Doumas, directeur des chemins de fer Pirée-Athènes-Peloponnèse, et Spyridon Katsoulidis (ECAM 1901), né en Roumanie, travaillent dans les chemins de fer en Grèce, alors que le crétois Nikolaos Petasis (EPC 1888) et Victor Stavros (ECAM 1893), né à Vienne mais dont l'adresse familiale pendant ses études en France est Constantinople, contribuent au développement du réseau ferré de l'Empire ottoman[30].

Une bonne minorité enfin des ingénieurs grecs ayant étudié dans les grandes écoles françaises, diplômée pour l'essentiel à la fin de la période étudiée ici, applique les connaissances acquises en France non pas dans les secteurs tradi-tionnels des travaux publiques et des mines mais au sein d'industries naissan-tes en Grèce au début du XXᵉ siècle[31]. L'électricité attire ainsi Konstantinos Argyropoulos (ECAM 1908), qui travaille de 1911 à 1914 pour la Compagnie Hellénique d'électricité[32]. Georgios Staphanopli (ECAM 1899) est ingénieur, de 1902 à 1910, de l'Énergie électrique du littoral Méditerranéen. Panagiotis Kok-kinopoulos (EPC 1892) fera aussi une carrière dans le secteur, d'abord comme ingénieur de la Compagnie Hellénique d'électricité (1903-1912), ensuite comme directeur d'une compagnie à la fondation de laquelle il participe. Evag-

---

30. Sur les chemins de fer ottomans, voir rapidement Donald Quataert, " The age of reforms, 1812-1914 ", in Halil Inalcik et Donald Quataert (eds), *An economic and social history of the Otto-man Empire, 1300-1914*, Cambridge, Cambridge University Press, 1994, (759-943), 804-815 en particulier.

31. Sur l'industrie grecque durant la fin du XIXᵉ et les premières décennies du XXᵉ siècle, voir : Stathis N. Tsotsoros, *I sygkrotisi tou viomichanikou kefalaiou stin Ellada, 1898-1939* (*La consti-tution du capital industriel en Grèce, 1898-1939*), Athènes, MIET, 1993 (2 tomes) ; Christos Had-ziiossif, *I giraia selini. I viomichania stin elliniki oikonomia* (*La vieille lune. L'industrie dans l'économie grecque*), Athènes, Themelio, 1993 ; Christina Agriantoni, *Oi aparches tis ekviomicha-nissis stin Ellada ton 19o aiona* (*Les débuts de l'industrialisation en Grèce au XIXᵉ siècle*), Athè-nes, Istoriko Archeio Emborikis Trapezas, 1986 ; Christina Agriantoni, " Viomichania (" L'industrie "), *Istoria tis Elladas tou 20ou aiona. Oi aparches : 1900-1922* (*Histoire de la Grèce au XXᵉ siècle. Les débuts : 1900-1922*), t. A, partie I, Athènes, Bibliorama, 2000, 173-221.

32. Sur le secteur d'électricité et ses professions, Nikos Pantelakis, *O eksilektrismos tis Ella-das. Apo tin idiotiki protovoulia sto kratiko monopolio, 1889-1956* (*L'électrification de la Grèce. Du capital privé au monopole d'État, 1889-1956*), Athènes, MIET, 1991 ; voir aussi Aliki Vaxeva-noglou, *I koinoniki ypodochi tis kainotomias* (*La réception sociale de l'innovation*), Athènes, Cen-tre de Recherches Néohelléniques, 1996.

gelos Patrinos (ECAM 1902), né en Russie, après une carrière en Russie, fonde en Grèce en 1920 un bureau d'études spécialisé dans les installations électro-mécaniques. Alexandros Stavridis (EP 1893 et EPC 1895), après avoir travaillé en France, fonde à Athènes un bureau d'études, qui œuvre, entre autres, dans le domaine de l'électricité. Dimosthenis Lykiardopoulos (ECAM 1900), né à Mersina (Empire ottoman), après une carrière dans sa ville natale, s'installe à Athènes où il dirige pendant plusieurs années (1919-1927) les usines de cons-truction mécanique et les chantiers de construction navale " Vasileiadis ". Nikolaos Chrysochoïdis (ECAM 1894), après avoir travaillé en France et en Tunisie, est employé à partir de 1924 comme ingénieur dans la firme de fabri-cation de ciments " Olympos " à Volos. Quant à Spilios (Spyridon) Agapitos (EPC 1896), il sera impliqué dans la fondation et la direction de plusieurs entre-prises de l'époque dont la firme de ciments " Iraklis ".

Si la majorité des élèves retournent, une fois le diplôme acquis, à leur terri-toire administratif d'origine — les Grecs de nationalité grecque au royaume grec, ceux qui sont sujets ottomans dans l'Empire, avant de s'installer, pour la plupart d'entre eux, en Grèce, quand la guerre éclate dans les années 1910 entre la Grèce et la Turquie —, des trajets cosmopolites ne sont pas rares. Spi-ridon Dendrinos, né à Constantinople, admis à l'EPC en 1876, démarre sa car-rière d'ingénieur à Istanbul en travaillant pour le compte de l'administration ottomane. Mais l'année suivante, en 1880, il offre ses services au gouverne-ment grec, avant de s'installer quelques années plus tard à Bucarest. M.I. Agelastos se trouve au début des années 1890 à Athènes comme ingénieur de la *Société internationale anonyme de construction et d'entreprises de tra-vaux publics*. Mais il ne tarde pas à traverser les frontières en direction de l'Orient pour travailler à Istanbul comme ingénieur à *la Société de chemins de fer ottoman d'Anatolie*. Dix ans plus tard, il retourne à Athènes mais pas pour longtemps puisque, en 1908, on le trouve impliqué dans la construction des chemins de fer en Russie. Dimitrios Papas (ECAM 1895), né à Istanbul, ira jusqu'en Indochine, avant de s'installer à Athènes.

Les élèves grecs dans la société civile et la vie politique

Nous avons vu que les ingénieurs grecs ayant étudié en France sont très pré-sents à la fois dans les administrations techniques de l'État grec, l'enseigne-ment supérieur du pays ainsi que dans les activités économiques à caractère technique (grands travaux, mines, électricité...) de la région (Grèce mais aussi Empire ottoman) tout au long du XIXe siècle et le début du XXe siècle.

Mais leur rôle dans la modernisation de la région ne s'arrête pas ici. Pour rester dans le cas grec que nous maîtrisons mieux, nos ingénieurs participent activement à ce que l'on pourrait appeler la modernisation " intellectuelle " et

" politique " du pays[33]. Modernisation " intellectuelle " d'abord. Plusieurs d'entre eux importent au pays des produits intellectuels élaborés en France. Impliqués dans la question de l'alimentation en eau d'Athènes à la fin du XIXe siècle, ils prêchent les idées et des pratiques du mouvement hygiéniste. Nombre d'entre eux (P. Protopapadakis et I. Aggelopoulos par exemple) se font également les avocats du municipalisme social, c'est-à-dire du mouvement en faveur de la gestion municipale de différents services techniques, celui des eaux et de l'assainissement en premier chef[34]. Themistoklis Charitakis (EM 1907), traducteur, en 1924, du théoricien français de la gestion rationnelle des entreprises Henri Fayol (1841-1925), se fait en Grèce le chantre du mouvement de rationalisation industrielle qui se développe dans les pays industrialisés pendant l'entre-deux-guerres. Le rôle d'intercesseur auprès de la société grecque de produits intellectuels venant de l'Occident assuré par nos ingénieurs se manifeste également dans le domaine plus abstrait des " styles de raisonnement " (I. Hacking), à savoir les diverses manières d'argumenter, de justifier, et de mettre à l'épreuve[35].

Modernisation " politique " de la société grecque ensuite. Plusieurs des ingénieurs formés en France jouent, en effet, un rôle de premier plan dans la création d'associations professionnelles indépendantes (organisations dites " horizontales ") dont les modes de fonctionnement et d'intervention vont à l'encontre des réseaux clientélistes bien implantés dans la société grecque de l'époque. Ils participent également à la constitution d'un espace et d'une opinion publics grâce à la création des périodiques et par l'intermédiaire d'interventions dans la presse de l'époque. Ainsi L. Vlassis, I. Aggelopoulos, P. Protopapadakis, I. Markopoulos, Dimitrios Aravantinos (EPC 1883) participent à la création, en 1898, de l'Association Polytechnique (Ellinikos Polytechnikos Syllogos), à savoir la première organisation du monde technique en Grèce. L'année suivante l'Association publie une importante revue technique au nom

33. Sur les thèmes de " modernisation intellectuelle " et de " modernisation politique ", traités à travers la question de l'alimentation en eau d'Athènes, voir Konstantinos Chatzis, " Le maire, le premier ministre et l'ingénieur : la difficile mise en place du réseau d'adduction d'eau à Athènes, 1830-1930 ", in Denis Bocquet et Samuel Fettah (éds), *Réseaux techniques et réseaux de pouvoir : la modernisation technique dans les villes européennes, 1850-1930*, Rome, École Française de Rome (à paraître).

34. Pour plus de développements, voir K. Chatzis, " Le maire, le premier ministre et l'ingénieur... ", *op. cit.*

35. Sur les différents styles de raisonnement développés par les ingénieurs aux XIXe et XXe siècles, et leur mise en œuvre dans les réseaux techniques et l'industrie (idéal analytique, idéal de l'automaticité, style de raisonnement statistique...), voir Konstantinos Chatzis, *La pluie, le métro et l'ingénieur. Contribution à l'histoire de l'assainissement et des transports urbains (XIXe-XXe siècles)*, Paris, L'Harmattan, 2000 ; K. Chatzis, " La fonction entretien durant les Trente Glorieuses ", *Revue Française de Gestion*, n° 135 (septembre-octobre 2001), 93-100 (Dossier : La gestion des ingénieurs au crible de l'histoire) ; K. Chatzis, " Searching for standards : French engineers and time and motion studies of industrial operations in the 1950s ", *History and Technology*, vol. 15, n° 3 (1999), 233-261.

d'*Archimidis*, d'une longévité assez importante (1899-1925) et dont la publication est assurée par I. Aggelopoulos, expert en la matière puisque quelques années auparavant il avait édité pendant deux ans (1887 et 1888) la *Revue Mécanique* (*Michaniki Epitheorisi*). Et quand, en 1923, l'actuelle organisation professionnelle des ingénieurs grecs, la Chambre Technique de Grèce (Techniko Epimelitirio tis Elladas) voit le jour, elle fait appel pour la présidence à I. Aggelopoulos alors que T. Charitakis occupe la fonction de Secrétaire spécial.

Nous ne pouvons pas terminer ce chapitre sans mentionner la participation d'un certain nombre d'ingénieurs passés par la France dans les évolutions politiques du pays. Dimitrios Soutsos (ECAM 1872) préside en tant que maire aux destinées de la ville d'Athènes pendant la période 1879-1887[36]. Ioannis Sechos, Phokion Negris (EP 1867 et EM 1869) et Linos Kogevinas (ECAM 1908) seront députés au parlement grec. V. Valtinos devient Ministre de la défense dans les gouvernements de Koumoundouros ; Ph. Negris effectue dans les années 1900-1920 des passages au Ministère de l'économie, de l'intérieur et celui des transports (créé en 1914). I. Argyropoulos et L. Kogevinas occupent au Ministère des transports le poste de ministre (1916-1917) et de ministre délégué (1932) respectivement. Mais la figure la plus connue, et la plus tragique, parmi ces ingénieurs qui entrent dans l'arène politique, est à coup sûr P. Protopapadakis. Député, ministre, il devient premier ministre en mai 1922. Jugé responsable de la défaite de l'armée grecque pendant la guerre d'Asie mineure, il est fusillé, avec cinq autres hommes d'État, le 15 novembre 1922.

CONCLUSION

Consacré aux élèves grecs dans les grandes écoles d'ingénieurs en France au XIXe siècle (École polytechnique, École des ponts et chaussées, École des mines et École centrale des arts et manufactures), le présent article procède d'un projet plus large en cours de réalisation qui a comme objet un groupe social particulier, les ingénieurs grecs[37]. Acteur dont l'importance pour la modernisation de la Grèce, mais aussi de la région plus large à laquelle celle-ci appartient (Empire ottoman, Balkans), semble être amplement attestée à travers plusieurs ouvrages récents, le monde des ingénieurs grecs, malgré l'existence de quelques travaux, est privé jusqu'à présent des analyses qu'il mérite. Tout en étant conscients du caractère d'ébauche que représente cet article qui procède souvent, faute d'une documentation suffisamment constituée, par des " touches impressionnistes ", nous espérons avoir au moins montré l'intérêt

---

36. Sur les actions de Soutsos en matière d'eau, voir K. Chatzis, " Le maire, le premier ministre et l'ingénieur... ", *op. cit.*

37. Nous plaidons ici en faveur d'une histoire sociale qui, sans nier l'intérêt et la nécessité des approches mobilisant de " grosses " entités telles que les " classes " ou " l'État ", veut restituer aux groupes intermédiaires le rôle qui leur revient dans les processus sociaux. Voir Christophe Charle, *Histoire sociale de la France au XIXe siècle*, Paris, Seuil, 1991.

que présente l'étude de la figure de l'ingénieur grec éduqué à l'étranger : figure polyvalente — la même personne travaille souvent alternativement pour le compte de l'État et celui du secteur privé, tout en participant à la mise en place d'un système éducatif moderne —, figure cosmopolite, à l'instar d'autres acteurs grecs de l'époque, tels que les négociants[38], d'un monde structuré en réseaux qui transpercent les frontières politiques[39].

Abréviations

EP 1850 : élève admis à l'École polytechnique en 1850.

EPC 1850 : élève admis à l'École des ponts et chaussées en 1850.

EM 1850 : élève admis à l'École des mines en 1850.

ECAM 1850 : élève diplômé de l'École centrale des arts et manufactures en 1850.

---

38. Sur la figure du négociant grec et ses mutations dans le temps, voir Georges Dertilis, " Entrepreneurs grecs : trois générations, 1770-1900 ", in Franco Angiolini et Daniel Roche (dir.), *Cultures et formations négociantes dans l'Europe moderne*, Paris, Editions de l'EHESS, 1995, 111-129.

39. Voir G. Dertilis (dir.), *Banquiers, usuriers et paysans. Réseaux de crédit et stratégies du capital en Grèce, 1780-1930*, Paris, La Découverte et Fondation des Treilles, 1988.

CROSSING COMMUNAL BOUNDARIES : TECHNOLOGY AND CULTURAL DIVERSITY IN THE 19th CENTURY OTTOMAN EMPIRE

Yakup BEKTAŞ

Reporting the ground-breaking for the Izmir-to-Aydin railroad in October 1857, the first railroad in the Ottoman Empire, *The Illustrated London News* observed the whole city of Izmir formed an " animated circle " around the field where the inauguration ceremony was to take place : Ottoman ministers, high ranking pashas, and civil and military authorities of Izmir, the Mufti or Muslim high priest, the Mullah or judge, the Greek and Armenian Bishops, the Great Rabbi of the Jews, the consuls, the general public[1]. The ceremony began with the Mufti offering prayers. Following this, Reshid Pasha, the Grand Vizier, saluted by artillery fire by the imperial troops, filled a small mahogany barrow with a silver spade, and emptied it at a spot designated for the purpose. Then, the consuls, and representatives of all the communities, repeated the action. A Muslim sacrifice of several sheep followed.

Half a year later, *The Times* (London) hailed the first whistle of the locomotive on this railway among the cheering crowds of as many as " nineteen races ", although it counted eighteen : " English, Irish, Scotch, French, Americans, Italians, Slovenians, Armenians, Turks, Greeks, Poles, Albanians, Austrians, Prussians, Hindus, Africans, Ionians, and Spaniards "[2]. The list reveals perhaps to some degree a British bias in enumerating the so-called races, no doubt because the reporter intended to glorify Britain's part in the undertaking. However, similar scenes were common in other parts of the empire in the aftermath of the Crimean War, when an Ottoman feeling in favor of technological progress and friendly relations with Europe began to develop. Every opening of a telegraph office, a railroad station, or a factory was an occasion for inter-

---

1. *The Illustrated London News* (31st October 1857), 436-437.
2. *The Times* (London) (6th April 1858).

communal celebration and solidarity. As in the vivid example of the inauguration of the Izmir-Aydın railroad, even the mere inauguration of novel public
facilities acted to draw people together.

The Ottoman Empire was ethnically and culturally a most diverse society. It
comprised communities of widely differing religious, ethnic, and linguistic
affiliations (called *millet*). The main division was into three religion-based
millets : Muslims, Christians and Jews. Each in turn comprised sub-communi-
ties, often ethnically and linguistically defined. Muslims, for example, included
the groups of Turks, Kurds, Arabs, Persians, and Europeans. Greeks and Armenians formed the largest Christian communities[3]. Although these communities
coexisted relatively peacefully for centuries, close social interactions between
them, especially among those of the three main religions, were traditionally
limited. Muslims, Christians, and Jews rarely mingled. Their dwelling quarters,
schools, and mosques, churches and synagogues, though often found side by
side, were a taboo for one another.

The religious and social boundaries that kept them aloof came to be questioned with the advent of technological and social change in the nineteenth
century. The expanding Ottoman cultural encounters and exchanges with Western Europe, especially in technological innovations and scientific practices created a meeting ground for interactions among these diverse communities. The
telegraphs, railroads, the applications of electricity, steam powered mills and
factories became foci of public attention as spectacles of the age, and raised
the interest of all Ottomans. These innovations not only aroused personal and
communal aspirations, but also created a forum for alliance and union.

Above all, these new technological enterprises employed and gave recognition to many members of non-Muslim communities, who had been largely
excluded from the high Ottoman circles. The participation of non-Muslim peoples in public works and institutions was sanctioned by the Tanzimat or
reforms of 1839, which recognized the equality of all peoples of the empire.
Consolidating this principle, the Islahat Fermanı or the reform charter of 1856
pledged equal treatment of all citizens in education, justice, taxation, military
service and employment in state institutions. The resulting social and political
reforms were responsible for the employment of a notable number of non-
Muslims in the central and local government institutions. This effort continued
as a state policy, especially during the reign of Abdulhamid II, to ensure social

---

3. For the religious and ethnic distribution of Ottoman populations see Kemal H. Karpat, *Ottoman Population, 1830-1914 : Demographic and Social Characteristics*, Madison, 1985, 45-59 ;
also *An Inquiry Into the Social Foundation of Nationalism in the Ottoman State : From Social
Estates to Classes, from Millets to Nations*, Princeton, 1973.

justice among the *millets*, while at same time promoting the empire's image in the West as a tolerant multiethnic society[4].

However, these reforms were perhaps only partially responsible for the disproportionately high representation of non-Muslims in the new technological and scientific enterprises. Because of their relative familiarity with European languages and cultures, the members of Christian communities, mainly Greeks and Armenians, were in the best position to take advantage of the new opportunities[5]. The long tradition of providing the empire with dragomans, and tradesmen and merchants, pushed these communities into closer communication with Europe. Their wealthy members had been able to send their children to colleges in Italy, France, Austria and Britain since the 18th century. They did so in greater numbers in the 19th century, and the educated students returned to the Ottoman Empire to meet the huge demand for Western skills and expertise generated by the new reforms. These individuals dominated the teaching and practice of modern medicine, and made up the overwhelming majority of Ottoman physicians and chemists in the late nineteenth century[6]. Furthermore, Greek and Armenian schools generally offered foreign languages, principally French. Another significant source of Western learning in the Ottoman communities was European and American missionary schools, established in large numbers from the 1830s throughout the empire. As Christians, Greeks and Armenians were in a better position to attend these schools than Muslims and Jews, who denounced them initially[7]. Greeks, and especially Armenians, had also the additional advantage of speaking the Turkish beside European languages[8].

Ottoman Greeks, in spite of keeping a low profile following the Greek revolution in 1821, became a major force in Ottoman modernization, particularly in light industries and textiles. Greek entrepreneurs set up cotton plantations,

4. On the Ottoman image building policies during the reign of Abdulhamid II, see Selim Deringil, *The Well-Protected Domains : Ideology and the Legitimation of Power in the Ottoman Empire, 1876-1909*, London, 1998.

5. For a discussion on the changing role of the non-Muslim Ottoman communities see Roderic Davison, " The Millets as Agents of Change in the Nineteenth-Century Ottoman Empire ", in Braude and Bernard Lewis (eds), *Christians and Jews in the Ottoman Empire : The Functioning of a Plural Society*, I, New York, London, 1982, 319-337 (2 vols), and Charles Issawi, " The Transformation of the Economic Position of the Millets in the Nineteenth Century ", *loc., cit.*, I, 261-272 ; Donald Quataert, *Manufacturing and Technology Transfer in the Ottoman Empire, 1800-1914*, Istanbul, 1992.

6. Ch. Issawi, " The Transformation of the Economic Position of the Millets in the Nineteenth Century ", *op. cit.*

7. An example is the seminary run by Cyrus Hamlin, a Congregational missionary who went to Istanbul in 1837. Hamlin later founded Robert College. Among his students were a significant number of Armenians. See Cyrus Hamlin, *My Life and Times*, Boston, Chicago, 1893 ; see also E. D.G. Prime, *Forty Years in the Turkish Empire : Memoirs of Rev. William Goodell*, New York, 1876.

8. On the significance of language ability as a prime qualification for Ottoman officer of the Foreign Ministry see Carter V. Findley, *Ottoman Civil Officialdom*, Princeton, 1989.

and built Western style factories to spin the cotton[9]. In Western Turkey, they ran mines and mining firms that exported minerals such as emery and chrome[10]. In Izmir, the home of a large Greek population, by the early 20th century they operated ninety percent of such industries as steam flourmills[11]. The Greek colonies in Italy, the Balkans, Russia, Egypt and Western Europe were not only the source of new ideas for Ottoman Greeks, but also placed them in an advantageous position for trading and commerce. They established hundreds of firms in major European industrial and trade centers. For example, about 1870, in the flourishing industrial and commercial British town of Manchester alone there were more than 160 Greek import and export firms[12]. Furthermore, a considerable number of Greeks, together with Armenians, worked as agents for European firms. The role of these communities in the development of Ottoman trade and commerce, a crucial channel for cultural and technical exchange, is not a matter for this paper, and has been amply discussed elsewhere.

Armenians participated in the introduction of Western innovations to the Ottoman Empire even more actively. Besides providing the government with most of its dragomans following the Greek revolution, they staffed such fields as the Mint (Darphane) and industrial works. During the reign of Abdülmecid (1839-1861), the brothers Ohannes Dadian (1798-1869) and Bogos Dadian (1800-1863), who came from a prominent Armenian family long in charge of the Baruthane or the Imperial Powder Works, directed most of the government's technological and industrial works, including Western-style factories, the sultan's model farm and geological explorations[13]. Ohannes Dadian was responsible in the 1820s for improving or inventing a device used in the manufacture of rifles. In 1835, Sultan Mahmud II sent him to Europe to study the industrial progress there. On his return, Ohannes set out to modernize the Imperial Powder Works by introducing new machinery from Europe. In 1840 he headed an Ottoman commission to Britain to make plans for introducing " useful arts and manufactures " to the country[14]. He made contacts with engi-

9. On the Ottoman manufacturing see D. Quataert, *Manufacturing and Technology Transfer in the Ottoman Empire, 1800-1914, op. cit.*, 7-8.

10. Elena Frangakis-Syrett, " The Economic Activities of the Greek Community of Izmir in the Second Half of the Nineteenth and Early Twentieth Centuries ", in Dimitri Gondicas and Charles Issawi (eds), *Ottoman Greeks in the Age of Nationalism : Politics, Economy, and Society in the Nineteenth Century*, Princeton, 1999, 17-44.

11. *Idem*, 32.

12. Charles Issawi, " Introduction ", in Dimitri Gondicas and Charles Issawi (eds), *Ottoman Greeks in the Age of Nationalism : Politics, Economy, and Society in the Nineteenth Century*, Princeton, 1999, 5.

13. For a brief history of the Dadians see Y.G. Çark, *Türk Devleti Hizmetinde Ermeniler, 1853-1953*, Istanbul, 1953, 75-9.

14. On this mission see *Minutes of Proceedings of the Institution of Civil Engineers*, 2 (1843) : 125-126 ; *The Life of Sir William Fairbairn, Bart* (Partly written by himself, edited and completed by William Pole), London, 1877, 165-176.

neers and entrepreneurs for developing industrial works in the Ottoman Empire. Through these efforts came some of the most ambitious Ottoman industrial projects. They included print works, cotton mills, cloth factories, mohair finishing plants, and silk manufacture in Istanbul, Izmit, Bursa and Izmir[15]. The Imperial Powder Works was expanded. Iron casting foundries were set up. Dozens of prominent British engineers, and hundreds of skilled workers were engaged for these works in the 1840s. Although most of the projects had relatively little success, they laid the foundation of these crucial 19th industries in the Ottoman Empire.

The Dadians also directed other Western style projects, introducing agricultural reforms and mining enterprises. One of these was the first American scientific mission to the Ottoman Empire (1846-1850). In 1845, upon Sultan Abdülmecid's request for American cotton planters, the United States government sent cotton experts, James Bolton Davis and John Lawrence Smith[16]. Davis set up a model farm with a school for agriculture in a village near Istanbul (today, Yeşilköy-Halkalı area), where the Dadians themselves owned a big farm. But Smith soon left to explore the mineral resources of the country. He discovered new ores in Western Turkey, of which those of emery became particularly lucrative. At this time, Ohannes hold one of the most prestigious Ottoman positions : " the Royal Director of the Powder Works of the Ottoman Government and Chief Inspector of the Imperial Factories "[17]. Ohannes was perhaps the first Ottoman to be elected as an associate member of the Institute of Civil Engineers in London, and later of other European engineering societies in Edinburgh and Paris[18]. The Dadians were of course not the only Armenians in prominent positions. Many others obtained top positions in the Public Works Department, particularly in engineering, agriculture, and the telegraph service.

The situation of the Jews, the largest non-Muslim non-Christian Ottoman community, presents a different picture. Although Jews had a considerable influence in reviving Ottoman medicine and sciences in earlier centuries[19], their share in the technical fields in the nineteenth century was relatively

15. On the extent of these works see, for example, US National Archives, Record Group 84, Washington, DC ; Charles MacFarlane, *Kismet*, London, 1853, 320-340.

16. US National Archives, Record Group 84, Washington, DC.

17. See Ohannes Dadian's testimonial letter to John Lawrence Smith, December 1849. The original is in Armenian. Archives of the Southern Baptist Theological Seminary, Kentucky.

18. Ohannes Dadian is described as an associate member of the Institute of Civil Engineers in London in *Minutes of Proceedings of the Institution of Civil Engineers*, 2 (1843), 125-126.

19. See Ekmeleddin Ihsanoğlu, " Some Notes on Andalusian Contribution to Ottoman Science in its 500th Anniversary ", paper presented at the 33th International Congress for the History of Medicine, 1-6 Sept. 1992, Granada, 6 ; *idem*, " Travel to the East in its Quincentenary : the Andalusian Contribution to Ottoman Science ", lecture delivered at University of California Los Angeles (UCLA), 10-17 Oct. 1992, California, 9.

small[20]. Historians such as Stanford Shaw explain this loss of influence by pointing out that Ottoman Jews had lost touch with Western learning and technology by the nineteenth century[21]. Jewish schools, for example, until late in the century did not include the study of Western languages. Seeing it as a threat to their authority, Jewish religious leaders initially opposed Ottoman modernization and denied their students modern education. As a result, Ottoman Jews began to take advantage of Western technologies relatively late[22]. In spite of this disadvantageous beginning, Jewish owned manufacturing businesses and light industries prospered in some towns such as Salonica, a prominent Ottoman province and one of the centers where Jews were particularly active[23]. On the other hand, prominent Jewish financiers and capitalists in Europe, such as the Rothschilds and Baron Maurice de Hirsch, invested heavily in industrial projects such as railways and shipping in the Ottoman Empire, in which they employed many of their co-religionists as agents and staff.

The Ottoman telegraph service represents a most international and diverse field, in which virtually all the empire's peoples were represented. It was perhaps the only Western innovation and institution which became established truly as Ottoman in the nineteenth century, while railroads and industrial works remained mostly funded and operated by Europeans. Ottomans from the very beginning owned and operated the telegraph system, and later manufactured most of the machinery and equipment needed to run it themselves. It was also truly Ottoman in its representation of its many peoples. A large number of Armenians, a notable number of Greeks, other Christians and also later Jews not only found opportunities for employment, but also obtained top positions and public recognition. As historians have noticed their number was disproportionately higher than that of Muslims[24]. This was particularly striking in the early period of the telegraph, when there were not many members of Muslim communities trained in European languages. The knowledge of languages was a crucial qualification for work with the telegraph, and the Ottoman system had to cope with more than a dozen languages within its domains. French was the main international language of the system, and the employees were required to speak it[25]. Early Muslim employees therefore came from the institutions such

20. On the role of Jews in the Ottoman and Islamic scientific world in the earlier times see Bernard Lewis, *Muslim Discovery of Europe*, London, 1982.

21. See Stanford J. Shaw, *The Jews of the Ottoman Empire and the Turkish Republic*, New York, 1991, 157-159.

22. *Ibidem*.

23. See D. Quataert, *Manufacturing and Technology Transfer in the Ottoman Empire, 1800-1914, op. cit.*, 7.

24. Roderic H. Davison, *Essays in Ottoman and Turkish History, 1774-1923 : The Impact of the West*, Austin, 1990, 152.

25. On the knowledge of languages as a prime qualification for employment in the Ottoman foreign ministry see Carter V. Findley, *Ottoman Civil Officialdom*, Princeton, 1989.

as the Translation Bureau, the main Ottoman institution that taught French and other languages. English became a major telegraph language in 1864 with the opening of the Istanbul to Fao line as a part of the British Indo-European communication system, because the Ottoman government had to employ clerks and operators with knowledge of English in the major stations[26]. Armenians, Greeks and other non-Muslim Ottomans, and foreigners overwhelmingly occupied the positions involving foreign services and languages, and mechanical skills. For example, a listing for three main telegraph offices (Central, Beyoğlu and Üsküdar) in Istanbul in 1861 shows that all such positions were held exclusively by non-Muslims[27]. But non-Muslims were by no means limited to these positions. Armenians, especially, were able to obtain the highest positions, such as General Directors and later Ministers. Between 1860 and 1870 most of the Ottoman general telegraph directors, for example, were Armenians : Davud Effendi (1860-1861), Franko Effendi (1861), Dikran Effendi (1862-1863), Aleko Effendi (1863), Agaton Effendi (1864-1868)[28]. (Agaton (Kirkor) and Davud also became Minister of Public Works in 1868 and 1868-71 respectively.) After the telegraph and postal services had merged in 1871, there were also Armenian ministers for the Post and Telegraphs. At the telegraph offices throughout the empire, Armenians and Greeks served as chiefs, superintendents, mechanics, linesmen, foremen, clerks and messengers[29]. At Beyoğlu, the empire's busiest telegraph office, for example, the non-Muslim directors included a Briton, a Frenchman, a Greek and two Armenians. The Greek director Antoniyadis was in office for 30 years[30].

The Ottoman telegraph office represented a unique space for inter-cultural experimentation. In an office in the interior in the 1870s, it was likely to find an Armenian director, a Turkish inspector, a Greek mechanic, and a British or French operator. Western travelers in the Ottoman Empire in the late nineteenth

26. Yakup Bektaş, " The Sultan's Messenger : Cultural Constructions of Ottoman Telegraphy, 1847-1880", *Technology and Culture*, 41 (2000), 669-696.

27. Nesimi Yazıcı's list for that year gives only one Muslim mechanic (Süleyman Effendi), but none in the foreign services among 82 employees. See Nesimi Yazıcı, " Tanzimat Döneminde Osmanlı Haberleşme Kurumu ", in Hakki Dursun Yıldız (ed.), *150. Yılında Tanzimat*, Ankara, 1992, 139-209 (190-192).

28. Nesimi Yazıcı, " Tanzimat Döneminde Osmanlı Posta Örgütü ", *Tanzimat'tan Cumhuriyet'e Turkiye Ansiklopedisi*, 6 (1652) ; Asaf Tanrıkut, *Türkiye Posta ve Telefon Tarihi ve Teşkilat Mevzuatı*, Ankara, 1984, 572.

29. There is not an inclusive list of Armenians and others in the telegraph service. Some of the prominent names are included in A. Tanrıkut, *Türkiye Posta ve Telefon Tarihi ve Teşkilat Mevzuatı, op. cit.* For some Armenians in the telegraph service in the eastern Ottoman provinces see Mesrob K. Krikorian, *Armenians in the Service of the Ottoman Empire, 1860-1908*, London, 1977 ; Çark gives the portraits of prominent Armenians in the Ottoman service, including the telegraph. See Y.G. Çark, *Türk Devleti Hizmetinde Ermeniler, 1853-1953, op. cit.* ; also see Sadi Koçaş, *Tarih Boyunca Ermeniler ve Türk-Ermeni İlişkileri*, Ankara, 1967.

30. For a list of chief officers in the Ottoman telegraph system see A. Tanrıkut, *Türkiye Posta ve Telgraf Teşkilat ve Mevzuati*, Ankara, 1984.

century were often surprised to see the individuals of different communities at work at the telegraph offices that dotted the empire[31].

This diversity later extended to shipyards, factories, and railroads and railroad stations. Railroads in particular offered greater employment opportunities for the members of non-Muslim communities[32]. Unlike the telegraphs, most Ottoman railroads were financed and operated by European companies. Their managers often brought from home their most skilled engineers and workers, and hired laborers of many other nations besides the Ottoman subjects. Among their favorite employees were Armenians, Greeks, and Jews, who worked as stationmasters, mechanics and locomotive drivers[33].

CONCLUSION

New Western technologies in the nineteenth century facilitated communication across the Ottoman Empire's cultural and ethnic boundaries. The social stimuli of new machinery, skilled men and Western instructors that urged the Ottoman world into contact with the West also forced the empire's previously segregated communities to mingle and interact. The traditionally separate realms of Muslims, Christians, Jews and other communities met at the telegraph office, railway station, passenger train, classroom, factory, and cotton field. These technological spaces offered numerous opportunities for contacts and communication, and helped change the dynamics of Ottoman communal structure. The resulting social change was not always harmonious, but sometimes caused subtle conflicts, jealousy, and perhaps was even partly responsible for some of ethnic tension especially from the late nineteenth century. This aspect of the process needs further study.

As far as employment qualifications are concerned, it appears that the Westernising Ottoman technical world was largely a meritocracy, which was partly shaped by the egalitarianism of the Tanzimat. For example, the prime reason for the high representation of Christian communities seems to be their knowledge of Western languages and know-how in the technical fields. But as the century wore on, more Muslims learned these languages and skills, and their share in the technical fields increased at the expense of the non-Muslims.

31. See, for example, Fred Burnaby, *On Horseback Through Asia Minor*, I, London, 1877, 70, 288 (2 vols) ; E.J. Davis, *Life in Asiatic Turkey*, London, 1879, 190, 471-472.

32. The ethnic diversity of workforce on Ottoman railroads has been the subject of some recent work. See the article by Peter Mentzel in this volume ; D. Quataert, *Social Disintegration and Popular Resistance in the Ottoman Empire, 1881-1908*, New York, London, 71-93.

33. Krikorian lists dozens of Armenians at such positions in Adana, Mersin and Tarsus alone. See Mesrob K. Krikorian, *Armenians in the Service of the Ottoman Empire, 1860-1908*, *op. cit.*, 66.

Were there other cultural factors at work besides the knowledge of Western languages and skills that might account for the notable presence of Christian Ottomans in the Ottoman technological enterprises ?[34]. Did, for example, Ottoman Christians feel a certain affinity for Western technologies, which they could consider essentially Christian ? Certainly there was a strong expectation by Western Europeans that their technologies would Christianize the empire. They claimed moral and intellectual superiority over non-Christian peoples. And they often favored their Christian brethren in the Ottoman Empire. Muslims and Jews, therefore, had reasons to be skeptical of European technological and cultural penetration. Unfortunately, there are not yet available extensive statistics of the Ottoman religious and ethnic communities that were involved in the new technological and industrial enterprises in the nineteenth century. A bigger and more accurate picture will surely be formed after such statistics are carefully analyzed. A balanced account of Ottoman technological experience in the nineteenth century could not be complete without giving these diverse communities their proper roles.

---

34. An example is the fact that Muslim women were initially not allowed working in Western style factories, which used mainly female labor. In 1872, in Bursa, an Ottoman silk and cotton industry center, Armenian and Greek women made up 95 percent of the workforce in the silk-reeling plants. See Ch. Issawi, " Introduction ", *op. cit.*, 5.

UNITY AND DIVERSITY ON OTTOMAN RAILWAYS : A PRELIMINARY REPORT ON TECHNOLOGY TRANSFER AND RAILWAY WORKERS IN THE OTTOMAN EMPIRE

Peter MENTZEL

A memo of the Oriental Railway Company reviewing personnel issues, briefly noted that four apprentices in the Traffic Division (" Verkehrseleven ") were promoted to Traffic Officers, Third Class (" Verkehrsbeamten III. Kl. ") on 1 July 1909. The fortunate apprentices, named Jacob Campeas, Israel Levy, Marc Rousseau, and Diomede Yatropoulos, were to receive salaries of 650 Kurus per month and free apartments[1]. This tantalizing glimpse of the life of the workers on one of the most important railroads of the Ottoman Empire raises a number of interesting questions connected with the general themes of occupational mobility and technology transfer. In particular, the promotion of these four men prompts us to wonder who these successful apprentices were, where they came from, and how they were hired and trained[2].

These questions can be unpacked and rephrased to produce three other questions. First, was there any interest on the part of the railroad companies, or the non-Ottoman employees of those companies, in passing on railroading skills to the Ottoman workers ? Second, was there any interest among Ottoman subjects in learning these skills ? Finally, what were the actual mechanisms for the transmission of these skills ? This essay represents a start in addressing those and other questions regarding railroad workers and technology transfer.

This paper will address these questions by examining four potential avenues by which railroading skills were transmitted to Ottoman workers. These include formal educational efforts organized by the railroad companies them-

---

1. Archives of the German Bank, 4.3.3.Sig.8003. " Personalien ", 97.

2. These and similar questions have been posed, though not intensively investigated, before. See, for example, Donald Quataert, *Manufacturing and Technology Transfer in the Ottoman Empire, 1800-1914*, Istanbul, Isis Press, 1992, x-xi.

selves or by the Ottoman state, private educational foundations or facilities, and, finally, informal or semi-formal "on-the-job-training" of railroad employees. The paper will argue that much of the skill transmission must have occurred on the job during interactions between Ottoman and non-Ottoman workers.

If this turns out to be the case, then it forces us to re-evaluate the way in which we have usually thought about foreign enterprises, especially railroads, in the Ottoman Empire. In particular, the stratified, divided, and segmented culture of work in which non-Ottoman workers represented a kind of labor aristocracy separate from their Ottoman co-workers would be in need of revision. Likewise, if Ottoman subjects employed by the railroad companies appear to have held a wide variety of jobs, it might indicate that the management of the companies was willing to share most of the technical skills involved with running a railroad with their Ottoman employees. Before beginning this investigation, it is important to note that this essay will focus almost exclusively on the workers who actually operated the trains, maintained the tracks and worked in the company offices. The construction workers who surveyed the routes, dug out the grades and tunnels, and laid the railroad lines, will not appear in this study. As other examinations of railroad labor have noted, construction workers, while they may have been engaged in building railroad tracks or buildings, occupy a different organizational space from the workers who were actually engaged in the operation of the trains[3].

THE RAILROAD COMPANIES

By 1914 there were (depending on how one counts) nine railroad companies operating 8,334 km of track on Ottoman soil[4]. All but one of these (the Hijaz Railroad) were European enterprises. That is, they were financed overwhelmingly by European capital. This is not to say that they did not include Ottoman shareholders, but only that most of the biggest shareholders and principal administrators were subjects of European countries. The railroad companies tended to be associated with the national interests of a particular European state, but they were all international in the sense that their shareholders were citizens of a wide variety of European and non-European states. Thus, while the Ottoman Anatolian Railway (generally known by the initials of its French name ; CFOA) was famously controlled by German financial interests, non-Ger-

---

3. See especially Walter Licht, *Working for the Railroad : The Organization of Work in the Nineteenth Century*, Princeton, Princeton University Press, 1983, xvii

4. The railroad companies operating on Ottoman territory in 1914 were the : Izmir -Aydin, and Izmir-Kasaba Railways ; Hijaz Railway ; Ottoman Anatolian Railroad (CFOA) ; Bagdad Railway ; Oriental Railway ; Mersin-Adana Railway ; Damascus, Hama, and Extension Railway (DHP), and the Jaffa-Jerusalem Railway. Vedat Eldem, *Osmanlı Imparatorluğu'nun Iktisadi Şartları Hakkında Bir Tetkik*, Ankara, Türk Tarih Kurumu Basımevi, 1994, 102-105.

mans had some important roles in the upper echelons of the company's administration[5].

By 1914, the railroad companies collectively employed 10,000 to 15,000 workers, both Ottoman subjects and foreigners[6]. In all of the companies, the Ottoman workers outnumbered the foreigners by wide margins. One scholar asserted that 90 percent of the workers on the CFOA were Ottomans[7]. A CFOA roster of the 670 full time, salaried employees (*i.e.*, not the entire workforce) listed 68 percent of the workers as Ottoman[8].

Within the workforce itself many have remarked upon the existence of an " ethnic division of labor ". According to this hypothesis, the upper levels of the companies were dominated by foreigners, the middle echelons by Ottoman Christians, and the lowest rungs of the ladder by Ottoman Muslims[9]. While this seems to have been broadly the case, the data in fact show that below the highest levels of railroad administration (which were indeed dominated by foreigners) there was a remarkably rich mixture of foreigners with Ottoman Christians, Jews, and Muslims. It was precisely this conglomeration of different ethnic groups and nationalities that made possible the transfer of skills.

While the railroads were constructed by European companies they had important connections to the Ottoman state. They operated under concessions granted by the Ottoman government which granted them the exclusive right to build and operate railroads in a particular part of the Empire's territory. These concessions were almost always accompanied by a " kilometric guarantee ". This was a kind of subsidy paid to the railroad company by the Ottoman state. It guaranteed the company a minimum income per kilometer of track laid. If the company's yearly net receipts fell short of this guarantee, the Ottoman treasury would make up the difference[10].

5. Of the 15 members of the CFOA's administration council (*Verwaltungsrat*) five were from Austria, France, or Switzerland. Archives of the German Bank, 4.3.3., Sig.8030. *Anatolische Eisenbahn Geselschaft/Société du Chemin de Fer Ottoman d'Anatolie. Bericht des Verwaltungsrates uber die Einundzwanziges Geschaftsjahr 1909*, 5.

6. Halil Inalcik and Donald Quataert (eds), *An Economic and Social History of the Ottoman Empire, 1300-1914*, Cambridge, Cambridge University Press, 1994, 810.

7. Donald Quataert, " A Provisional Report concerning the Impact of European Capital on Ottoman Port and Railway Workers, 1888-1909 ", in Jean-Louis Bacque-Grammont and Paul Dumont (eds), *Économie et Sociétés dans l'Empire Ottoman*, Paris, Centre National de la Recherche Scientifique, 1983, 467.

8. Archives of the German Bank, 4.3.3. Sig.8049, " *Personnel Commissionne* ".

9. H. Inalcik and D. Quataert (eds), *An Economic and Social History of the Ottoman Empire, 1300-1914, op. cit.*, 810-811. Basil C. Gounaris, *Steam over Macedonia, 1870-1912*, Boulder, East European Monographs, 1993, 68. D. Quataert, " Labor and Working Class History during the Late Ottoman Period ", *TSA Bulletin*, XV, 2 (1991), 370.

10. For general descriptions of the kilometric guarantee system, see H. Inalcik and D. Quataert (eds), *An Economic and Social History of the Ottoman Empire, 1300-1914, op. cit.*, 806-807.

There has been a significant amount of controversy in the literature regarding the nature and effect of the kilometric guarantees. Some have seen it as a blatant example of European economic imperialism while others have downplayed its significance. The former point to the shameless exploitation of the system by such financiers as Baron Maurice Hirsch who reaped enormous profits by building shoddy and non-contiguous bits of railroad in the Ottoman Balkans. Others point to the (allegedly) haphazard and meandering routes rail lines took to maximize the amount of track laid[11]. Others have noted that Hirsch's antics seem to have been exceptional and that the routes the railroads took were usually rational and frequently dictated, at least in part, by the Ottoman government itself[12]. It could also be pointed out that subsidies to railroad companies were certainly nothing unusual and were granted by most governments.

RAILROAD COMPANIES AND TECHNOLOGY TRANSFER

Whether or not railroad companies were interested in using the kilometric guarantee as a means of bilking the Ottoman treasury out of its money, it is very clear that the companies had no interest in fostering Ottoman industrial development. The railroads' receipts were almost exclusively from the transport of passengers or raw materials, especially agricultural products such as wheat, figs, nuts, etc. The companies imported all of their machinery from Europe and made no attempts to help establish factories or workshops in the Ottoman Empire to manufacture any of the equipment or tools needed to operate the railroads. Thus, not only was the rolling stock imported from Europe, but ties (or " sleepers " in British English), rails and sometimes even coal, were all imported[13].

This development stood in contrast to what many contemporary observers (including Karl Marx), believed would happen. They argued that the increase in railroad construction in the Empire would eventually lead to the establishment of the industries needed to provide the rolling stock and machinery needed to operate and maintain the trains and tracks[14]. The fact that this did not happen is almost certainly due primarily to the lack of interest on the part of the railroad companies to the long-term economic development of the Ottoman Empire. It also reflects, however, the inability or unwillingness of the

11. For example, Nicholas Fatih, *The World the Railways Made*, New York, Carroll and Graf Publishers, 1991, 170-173.

12. Charles Issawi, *The Economic History of Turkey, 1800-1914*, Chicago, University of Chicago Press, 1980, 149-150. John Kolars and Henry J. Malin, " Population and Accessibility : An Analysis of Turkish Railroads ", *The Geographical Review*, LX, 2 (1970), 240-241.

13. H. Inalcik and D. Quataert (eds), *An Economic and Social History of the Ottoman Empire, 1300-1914*, *op. cit.*, 813.

14. Ch. Issawi, *The Economic History of Turkey, 1800-1914*, *op. cit.*, 148.

Ottoman state itself to encourage such industrialization. As Quataert and others have made clear, state encouragement of industrialization was hampered because of the inability of the Ottoman government to use tariff barriers for such a purpose.

While the railroads were thus not interested in acting as agents of technology transfer in general or as the pioneers of the industrialization of the Ottoman Empire, they did work to diffuse skills necessary for the operation of the railroads to their employees and workers.

THE ORGANIZATION OF THE WORKFORCE

On the CFOA and the Oriental Railway Company lines, the workforce was divided into two broad categories, the *Agents* (or *Personnel*) *Commissionnés* and the *Agents en Régie,* corresponding roughly to salaried employees (the former) and hourly workers (the latter). The Agents Commissionnés represented the upper echelons of the workforce, while the Agents en Régie were made up of apprentices, day-laborers, etc. The Agents Commissionnés were all employed on contracts and drew salaries. They all had to participate in the Company retirement and health insurance fund. The Agents en Régie likewise had to contribute to the health insurance plan, but were not required to participate in the retirement account, although they could if they wished. The Agents Commissionnés also had available to them, on a case by case basis, company housing or supplementary funds to pay for housing[15].

On the Oriental Railway Company's lines, the Agents Commissionnés were themselves divided into four sections, each in turn divided into eight levels. The four sections corresponded to the different operational areas of a railroad, namely, " general services ", " movement and traffic ", " material and traction ", and " road and right-of-way ". The eight levels corresponded to the different levels of power and responsibility within each section. Thus the first level was always occupied by the section director, while the eighth was made up of the lowest paid jobs. For example, in the Traction division, the first level was held solely by the Chief Engineer. In the sixth level were locomotive drivers and Office Employees, Third Class. Some of the employees in the eighth (lowest) level were the conductors, office waiters, and assistant firemen[16].

All appointments to the ranks of the Agents Commissionnés had to be approved by the company directors and all employees had to pass examina-

15. German Bank, 4.3.3. Sig.8003. Betriebsgesellschaft der Orientalischen Eisenbahnen. " Reglement Concernant le Personnel ", 13, 14.

16. German Bank, 4.3.3. Sig.8003. B. der Orientalischen Eisenbahnen. " Eschelle des Traitements des Fonctionnes et Agents Commissionnes ".

tions. They also had to be between 18 and 30 years old, in good health, with normal vision and hearing. Finally, they had to be able to speak "a language appropriate to their occupation" (*langues en rapport avec le poste à lui confier...*)[17].

The *Agents en Régie* were made up mostly of day-laborers and apprentices. According to the Oriental Railway Company's personnel regulations, they included most of the ranks of the switchmen, brakemen, security guards, porters, flagmen, etc. Like the Agents Commissionnés, their appointments had to be confirmed by a section head and they too had a language requirement, although in their case they had to have a reading and writing knowledge of a local language (*une langue du pays*). On the other hand, unlike the Agents Commissionnes, they apparently needed only a rudimentary education (*une certaine instruction générale*)[18].

The different language requirements for the two different grades of employment are noteworthy. As most of the railroads in the Ottoman Empire used French as the language of administration, knowledge of that language would seem to have been necessary for a position as an Agent Commissionné in a managerial position. Some authors have argued that this situation effectively blocked Turkish speakers from most upper-level jobs[19].

Based on the information gleaned from Company sources, therefore, it seems that it was possible to rise through the ranks in this system, although the data have not yet revealed the upper limits of advancement. The Oriental Railway Company's personnel regulations, however, provide details on the system of advancement. For example, Station Apprentices had the opportunity to advance into "executive service" (*service exécutif*) by passing a series of exams at three month intervals (perhaps the four Traffic Apprentices who were mentioned in the introduction to this essay had to pass a similar set of tests). These exams tested their knowledge of telegraphy, of station operation, and of commercial service[20]. Similarly, locomotive firemen could gain promotion from the eighth to the seventh levels in the Agents Commissionnés hierarchy by passing an exam. The apprenticeship period on the Oriental Railways lasted four years until 1911. In that year, bowing to pressure from the railroad workers, the Oriental Railway Company administration lowered the apprenticeship period to three years[21].

17. *Idem*, "Reglement", 2-3.

18. *Idem*, 34-35.

19. Ziya Gürel, *Kurtuluş Savaşında Demiryolculuk*, Ankara, Türk Tarih Kurumu Basımevi, 1989, 13, 81.

20. German Bank, B. der Orientalishen Eisenbahnen, "Reglement", 4-5.

21. *Idem*, Gross to Verwaltungsrat, 151.

These bits of data strongly suggest that promotion within the Oriental Railway Company's hierarchy was indeed possible, but that at least some workers were unsatisfied with certain aspects of the Company's rules for advancement. Hence, the workers' demand for the shortening of the apprenticeship period.

FORMAL EDUCATION

The foregoing summary of the organization of the workforce on two main railroads indicates that railroad employees were expected to have some education or training and that employees could, at least theoretically, advance through the ranks based on passing certain examinations. What is much less clear is how these employees were to gain these skills.

There is some evidence that the railroad companies themselves and the Ottoman state provided at least some educational facilities for training students in elements of railroad work. In August 1872, for example, the state-run Sanayi Mektebi sent three students named Süleyman, Osman, and Ahmet (all apparently Muslims, judging by their names) to work in the Oriental Railway Company's machine shops at Yedikule, Istanbul. The chief engineer there, a French citizen named Lippman supervised apprentices studying iron work, carpentry, and locomotive repair, among other subjects[22]. The Ottoman government's Ministry of Endowments (*Evkaf*) also apparently ran a school of railroad work, at least in 1911. The graduates that year were placed with the CFOA, Baghdad, Izmir-Aydın, and Hijaz railroads. The curriculum included courses in train routing and movement, bookkeeping, telegraphy and railroad security[23].

Curiously, however, there is little additional information in the records of the railway companies or in the Ottoman government's archives detailing schools, seminars, or other organized technical training. If this paucity of information reflects the actual level of formal training for railroad workers, then it indicates that apprentices and other workers must have had some supplemental or alternative methods of instruction. In other words, workers must have had the opportunity to learn on the job from more experienced or better trained coworkers. But who were these workers and how did they interact ?

ON THE JOB TRAINING

The skilled workers on Ottoman railroads were, at least initially, non-Ottomans. Thus, if skills were indeed taught semi-formally while on the job, one would expect to find a situation where workers of different ethnic and national

22. Vahdettin Engin, *Rumeli Demiryollari*, Istanbul, Eren Yayıncılık, 1993, 78.
23. Osman Ergin, *Türkiye Maarif Tarihi*, vol. IV, Istanbul, Osmanbey Matbaası, 1942, 1307.

backgrounds worked in close proximity to each other and shared a friendly, or at least cordial, relationship. An examination of the culture of work and the backgrounds of workers on train crews of the CFOA and Oriental Railway Company lines argues strongly, if indirectly, for both of these conditions. Who were these workers and what were their origins ?

Unfortunately, we know little about the origins of the railroad workers, whether Ottoman or non-Ottoman. There is some anecdotal evidence that many of the Ottoman workers were locally recruited. In other words, they were hired directly from the communities through which the railroad passed. A list of the housing allowances and arrangements of Oriental Railway workers, for example, noted that many of them lived with their siblings, parents or in-laws[24]. This might be an indication that the workers themselves were natives of that particular town. Likewise, one scholar has speculated that the relatively large numbers of Greeks and Jews on the Salonika-Dedeağaç Junction Railroad reflected the general population of those groups in the region[25].

The non-Ottoman railroad workers were a truly international group. A CFOA employee roster of 670 salaried employees shows that the biggest number of non-Ottoman employees were Germans (seven percent of the total), followed by Austrians and/or Hungarians (six percent), Italians (five percent), and Greeks ([subjects of the Greek Kingdom], four percent). French and Swiss citizens each made up about three percent of the employees on the list. There were also Belgian, Bulgarian, English, Polish, Romanian, Russian, and Serbian employees, each representing one percent or less of the total employees[26].

As in the case of the Ottoman workers we unfortunately have very sparse information about these men. Dr. Archangelos Gabriel, an important labor organizer on the CFOA, claimed that most of the foreigners employed had no knowledge of the railroad business. However, reports and memos from the Oriental Railway Company archives show that at least some certainly did. For example, an Oriental Railway Company memo regarding personnel issues mentions a certain " Engineer Jokel " who came to the company from the Austrian Northwest Railroad. On the other end of the company hierarchy, a letter to the company directors dated 12 January 1912 discusses the employment of a Dr. Karl Wiedemann as a chief inspector of the railroad. Dr. Wiedemann, after what the letter describes as a solid career in transportation law, was employed by the Swiss Federal Railroads[27].

---

24. German Bank Archives, 4.3.3., Sig. 8003. B. der Orientalischen Eisenbahnen, " Releve : les indemnites de logement ", 40-41.

25. B.C. Gounaris, *Steam over Macedonia, 1870-1912, op. cit.,* 69.

26. German Bank Archives, 4.3.3., Sig.8049, " Personnel Commissioné ".

27. German Bank Archives, 4.3.3., Sig.8003, " Gehalts und Lohnaufbesserun ", 86. Gross an Verwaltungsrat, 12 January 1912, 177.

Foreign workers, no matter what their rank, were apparently enticed to work for the Ottoman railroad companies by the offer of higher wages than they could get at home[28]. Indeed, this wage difference was an important grievance among some Ottoman workers during the 1908 strike against the CFOA. The main organizer of that strike, Dr. Gabriel, pointed out that foreign workers earned much more than they could in their home countries[29]. The railroads also apparently made some effort to provide schools for the children of foreign workers. Thus, the Oriental Railway Company sponsored German language schools in Filibe (Plovdiv, Philippopolis) and Salonika, and the CFOA established German schools at Haydarpaşa and in Eskişehir[30].

The foreign workers held a wide range of jobs. As noted earlier in this paper, most scholarship has argued that the work force was divided, or segmented, along national and ethnic lines. While it seems broadly true that this was the case, and especially that non-Ottoman workers held the most important and well paid positions, there were exceptions to this general rule[31]. Indeed, while Ottoman subjects seem to have dominated the lower levels of employment, they were present in all types of positions, including some very important ones. On the other hand, non-Ottoman workers could be found, albeit in relatively small numbers, on the lowest rungs of the railroad employment ladder.

This mixture of Ottoman and non-Ottoman workers is crucial in understanding how skills could have been transferred. If these workers had the opportunity to work together in the same sorts of jobs it is highly likely that the skills necessary for these jobs were transmitted. During the first few years of railroad operation, most jobs must have been dominated by non-Ottomans, since Ottoman subjects would not have had the necessary skills. Yet, a diffusion of skills was taking place on the railroads since some of these jobs were filled (with some exceptions) by Ottomans after the railroads had been operating for several years.

One of the centers of the technological skills needed to operate the railroads were certainly the machine and maintenance shops. These were impressive establishments that employed hundreds of workers. The Salonika-Monastir

28. H. Inalcik and D. Quataert (eds), *An Economic and Social History of the Ottoman Empire, 1300-1914, op. cit.,* 811.

29. A[rchangelos] Gabriel, *Les Dessous de l'administration des Chemins de Fer Ottomans d'Anatolie et de Bagdad,* 134.

30. German Bank Archives, 4.3.3., Sig.8003, " Personalien ", 96. D. Quataert, *Social Disintegration and Popular Resistance in the Ottoman Empire, 1881-1908,* New York, New York University Press, 1983, 77. B.C. Gounaris, *Steam over Macedonia, 1870-1912, op. cit.,* 68.

31. A detailed discussion of the " ethnic division of labor " is more-or-less beyond the scope of this paper. I examine this phenomenon in depth in a forthcoming paper.

Railroad (part of the Oriental Railway Company's network), had 200 workers (machinists and laborers) in its Salonika repair shops[32]. The Eskisehir shops of the CFOA employed 420 workers, most of them Ottoman subjects[33]. As already noted, the Oriental Railway Company seemed to accept graduates of Ottoman technical schools as apprentices at its Yedikule repair shops, at least occasionally. Nevertheless, if the thin data on this subject are any indication, the bulk of the technical training Ottoman workers received apparently took place during their work in the shops.

An examination of the workforce in the machine shops is complicated by the lack of clear job descriptions for different classifications of workers. In the 1908 CFOA company roster referred to earlier in this essay, there are four job classifications that almost certainly were found in connection to the repair shops. These positions (and the number of employees within each position) were *agent technique* (28) ; *chef d'atelier* (1) ; *mécanicien des machines fixes* (2) ; and *chef monteur* (2). I did not include listings for *ingénieur* since an engineer could have conceivably been employed elsewhere in the company's operations ; in design, technical drawing, or drafting, for example. I could not find any job description for *agent technique* but the name strongly indicates that they had technical jobs and thus would almost certainly be found in the machine shops.

The breakdown of these jobs shows that the most important positions were held by foreigners. Thus, as in the Yedikule shops, the Chief of the Eskişehir machine shops was a German, Joseph Trub. Below him were the two head fitters (*chefs monteurs*). They were Andreas Burgis and Joseph Rauch, both Germans. In the next two classifications, the *agents techniques*, though more numerous, were on average better paid than the *mécanicien des machines fixes*. Of the latter, one each was Ottoman and Austrian. Finally, of the *agents techniques*, 11 were Germans ; five were French, four were Ottomans ; three were Swiss ; three were Italians, and one each were Romanian and Greek (*i.e.*, a subject of the Greek Kingdom, not an Ottoman Greek). Information for the Hijaz Railway's shops in Damascus reveal a similar breakdown of jobs and nationalities. The engineering and maintenance shops employed 163 men in 1909. Of these, 13 were non-Ottomans. Of those 13, the locomotive superintendent and a foreman were French, the works manager was British, and two electricians were Germans[34].

32. *Selanik Vilayeti Salnamesi, 1324*, 607.

33. D. Quataert, *Social Disintegration and Popular Resistance in the Ottoman Empire, 1881-1908, op. cit.,* 79.

34. William Ochsenwald, *The Hijaz Railroad*, Charlottesville, University Press of Virginia, 1980, 96.

Two points are immediately striking. One is the dominance of European, and especially German, personnel in the machine works. In the case of the CFOA, all of the overseer positions seem to have been held by Germans who are a plurality (though not, it should be emphasized, a majority), of the other workers. The second interesting point is that if we compare these figures with those for the overall employment in Eskişehir, it is apparent that most of the workers were not salaried employees. That is, they were *Agents en Régie*, which included apprentices.

Most of the salaried employees working as mechanics in the CFOA's shops therefore seems to have been non-Ottomans, although Ottomans were among the largest groups by nationality (overall, more numerous than either Austrian, Swiss, Italian, Romanian, or Greek). A further examination of the company's records indicates that the employment profile of the train crews was rather different. Among the ranks of locomotive drivers (" engineers " in American English) and firemen, Ottomans seem to have dominated. The same CFOA 1908 roster of salaried employees lists the names and ranks of 28 drivers along with their salaries and nationalities. 14 of these drivers were ranked First Class, 10 as Second Class, and one as Third Class. The remaining three were simply listed as drivers (*mécanicien*) with no class ranking. According to nationality, 23 of the drivers were described as " Ottoman ", two as " Austrian ", and one each as " English ", " French ", and " Italian ". The national profile of locomotive drivers on the Hijaz Railway was similar. In 1905, there were only two foreigners, both Germans, employed as locomotive drivers[35].

This data breaks down in interesting ways. First, it is obvious that Ottoman subjects overall made up the majority of the engine drivers. Thus, they also made up most of the highest rank. Of the five non-Ottomans, however, all were also First Class except for the English and French engineers, both of whom were described simply as *mécanicien*. Significantly, the only Third Class driver was an Ottoman named Ali Mehmed.

In the Oriental Railway Company, the locomotive drivers occupied the sixth level in the occupational hierarchy. Directly below them, in the seventh level, was the rank of *chauffeur examiné*. These were firemen who had apparently passed some sort of company examination or test. There were other firemen, in the eighth (and lowest) occupational level, who were not described as *examinés*. They might have held a status similar to that of apprentice. We do not know if the CFOA employed the same job hierarchy, but it too had both " examined " and unexamined firemen. The employee roster listed 28 examined firemen and one unexamined one. Of the examined firemen, 26 were

35. W. Ochsenwald, *The Hijaz Railroad, op. cit.* Ochsenwald comments that many of the Ottoman locomotive drivers were recruited from the naval dockyards.

described as Ottoman, and one each Italian and English. The unexamined fire-man was Greek (*i.e.*, a subject of the Greek kingdom). Thus, the overwhelming number of firemen were Ottoman subjects. On the Hijaz Railway, on the other hand, by 1914 "most firemen ... were French, Greek, or Lebanese Christians "[36].

A closer examination of the names of the drivers and firemen employed by the CFOA is also interesting. While it is admittedly hazardous to ascribe an eth-nic identity to a person on the basis of his name, names nevertheless can offer at least a very rough indicator of ethnicity. Hence, of the 23 drivers who were Ottoman subjects, 12 had Christian names while 11 had Muslim names, an almost even split. Similarly, of the 26 firemen who had passed their examina-tions, 11 had Christian names and 15 had Muslim ones, once again, a very nar-row division.

Thus, although non-Ottomans and Ottomans clearly dominated different aspects of work, both groups were represented in most work spaces. Workers of different nationalities and ethnic groups therefore worked side-by-side and thus had opportunities to learn from each other. The skilled non-Ottoman workers would have the opportunity to teach, or at least demonstrate, these skills to their Ottoman counterparts. Likewise, the workforce in the machine shops, especially the huge installation at Eskisehir, must have provided similar opportunities for the informal transmission of skills. But such opportunities would only have existed if the relationship between the workers of different ethnic and national backgrounds was conducive to friendship, or at least rudi-mentary cooperation.

RELATIONSHIPS AMONG THE WORKERS

The workers of different nationalities and ethnicities seem to have gotten along quite well. The hazardous nature of railroad work has been noted in other contexts for its fostering of a spirit of solidarity among railroad workers. It is possible that the nature of railroad work, with its various attendant haz-zards, fostered a spirit of camaraderie among all of the workers, regardless of their national or ethnic backgrounds. Data for the exact numbers of accidents per mile of track or number of train trips are not available, but there is much anecdotal evidence that Ottoman railroad men were plagued not only by acci-dents but by the attacks of bandits and terrorists as well. For example, in Mace-donia there were two bomb attacks against freight trains within two weeks of each other in October 1910. Both incidents caused serious damage. A case of

36. W. Ochsenwald, *The Hijaz Railroad, op. cit.* Ochsenwald does not provide the total num-ber of firemen (nor locomotive drivers) on the railroad.

sabotage involved the unscrewing of rails from a bridge spanning a river. The Orient Express very narrowly avoided a serious accident[37].

The railroad workers in the Ottoman Empire were also living in an intellectually exciting time, filled with ideas that would have fostered international and trans-ethnic solidarity. Among the Ottoman workers, the idea of Ottomanism might have exerted some influence, especially in the years immediately after the Young Turk Revolution. These workers might well have identified with their fellow workers, or the particular railroad company, as much as they identified with a national or ethnic group. Indeed, even during the Turkish War of Independence, characterized as it was by fierce inter-ethnic fighting, many Ottoman Greek and Armenian railroad employees not only remained in their jobs, but worked, often at great peril, to keep the trains running[38]. Feelings of class or trade solidarity are also evident in some of the railroad strikes of the Ottoman period. Workers of many different backgrounds were usually able to work together to bring the strikes to a more-or-less successful conclusion.

Again, the Oriental Railway Company's different lines give us many interesting glimpses and suggestions of this sort of trans-national and inter-ethnic cooperation. A glance at the names of the strike committee members hints at the rich ethnic mix of the strikers. The committee officers were named Yaglitziyan, Aidonides, Rotnagel, Melirytos, Lupovitz, and Diner. The other members of the committee were named Gibbon, Hatzopoulos, Eliades, Yeser, Gerke, Yovantsos, Paravantsos, Hussein, Romanos, and Blau[39]. In reports by the British Consul in Filibe and by the company officials to the Ottoman authorities other names appear. In particular, one of the most important of the railroad labor leaders was a certain Georg Rump. According to the above mentioned reports, his fellow " agitators " were named Boghos Hoschlian, Etienne Davidian, Athanasoff, Panayotoff, Koeff, and Theodorieff. Hoschlian and Rump appear in company records as salaried train-crew employees receiving a housing allowance. The British Consul's report describes Etienne Davidian as " an Armenian Ottoman subject and barrier guard dismissed last year for neglect of duty "[40]. Thus, these workers, though of different national and ethnic backgrounds, were clearly able to work (and strike) together.

The sorts of social interactions that these workers had are very difficult to reconstruct. Again, we must rely on hints and inferences. An intriguing glimpse of these sorts of contacts is provided in a report of living conditions

---

37. *Levant Herald*, 1910.

38. Z. Gürel, *Kurtuluş Savaşında Demiryolculuk, op. cit.,* 5, 22.

39. Alkiviades Panayotopoulos, " The Hellenic Contribution to the Ottoman Labor and Socialist Movement after 1908 ", *Études Balkaniques*, 1 (1980), 46.

40. BBA/BEO. TFR-1-A. 39/3852. F.O.371/552/35322 :82.

of some of the employees of the Oriental Railway Company. Of the 63 men listed, only nine inhabited their own house (*propre maison*). 12 are noted as being single while only eight are married (the other employees listed have no accompanying information on their marital status). Four live with brothers or in-laws and one Ezra Avigdor lived with his apprentices. The general impression that emerges from this glimpse of housing arrangements is not one of stable, married home life. A possible result of this domestic situation is alluded to in an Ottoman government report from the time of the 1908 railroad strikes. The report notes that workers in Filibe met in a pub (*birahane*) with strikers from the Sofia line (in Bulgaria)[41]. While it is impossible to prove, it is tempting to imagine that the workers chose to meet in the pub since it was a place with which they were familiar and comfortable, and which was beyond the gaze of the authorities.

CONCLUSIONS

This essay has provided some material with which we can offer some provisional answers to the questions posed in the introduction. The available data still do not allow us to answer the questions precisely but they at least provide us some broad clues.

First, the railroad companies plainly demanded that their employees, at least their salaried contract employees, be well trained. The Oriental Railway Company required school certificates of its contract employees and even day laborers had to be literate in a local language. Secondly, the same company apparently offered the opportunity of promotion to workers who could pass a series of exams. Unfortunately, we are still faced with the intriguing question of how these workers could study to pass these exams. I have found no indication that any of the railroad companies operated schools or provided classes for their employees, although this paper has presented evidence that alludes to apprenticeship programs of some kind on both the CFOA and Oriental Railway Company operations. In the absence of any hard evidence for extensive formal training programs, therefore, we can tentatively conclude that many of the railroading skills must have been taught and learned informally. That is, non-Ottoman skilled workers intentionally or unintentionally instructed their Ottoman coworkers.

The data presented in this paper show that both Ottoman and non-Ottoman subjects worked on the train crews and in the machine shops. According to at least the Oriental Railway Company's policy, all employees had to speak one of the local languages or a language necessary for his position. Thus, workers

41. BBA/BEO. A-Mtz-04. 159/28

of different national or ethnic backgrounds would likely have the means to communicate with each other. This paper has also argued that the relationship between workers of different national backgrounds seems to have been generally cordial, if not warm. The rigors and dangers of railroad work, as well as strikes against the companies, might have contributed to this comradery. In such a situation, it seems quite possible that apprentice workers would learn from European employees. Indeed, the early history of railroading in the USA offers an interesting mirror for this model of technology transfer. Some of the first locomotive engineers on American railroads were British and were frequently hired along with their locomotives. Over time, these foreign engineers taught their firemen or other train crew members how to operate and care for the locomotive[42]. A similar sort of dynamic inevitably arose on Ottoman railroads.

This model of technological skills transfer contrasts sharply with the situation on European-dominated railroads elsewhere in the non-Western world. The British railroads in India offer a striking counter example. Unlike the Ottoman case, British railroad operations in India were strictly segregated by ethnicity with the indigenous population permitted to learn only those skills thought appropriate by the railroad company. Particularly interesting in comparison to the Ottoman case were the ethnic restrictions for the train crews. The railroad companies hired European (mostly British) locomotive drivers exclusively until the turn of the century when they began to hire some Indian drivers for shunting or freight operations. Even so, as late as the 1930s express train drivers were almost exclusively European[43].

Thus, this preliminary investigation into the mechanisms for the transfer of railroading techniques and skills among Ottoman workers suggests not only what those mechanisms might have been, but also indicates something about the state of inter-ethnic and trans-national relations in the late Ottoman Empire.

---

42. W. Licht, *Working for the Railroad : The Organization of Work in the Nineteenth Century*, op. cit., 42.

43. Daniel R. Headrick, *The Tentacles of Progress*, New York, New York University Press, 1988, 322.

ORIENT ET OCCIDENT, LECTURES D'UNE POLARITÉ SCIENTIFIQUE ET TECHNIQUE

Antoine PICON

Dans un des manuscrits du *Bouvard et Pécuchet* de Flaubert, on peut lire la note suivante : " Bouvard voit l'avenir de l'humanité en beau. L'homme moderne est en progrès. L'Europe sera régénérée par l'Asie, la loi historique étant que la civilisation aille d'Orient en Occident (...). Les deux humanités enfin seront confondues[1] ". À l'image du héros de Flaubert, la culture française du XIXᵉ siècle s'est passionnée pour les rapports entre l'Orient et l'Occident, des rapports souvent interprétés sous l'égide d'une réconciliation, voire même d'une fusion à venir entre les deux aires culturelles aux destins longtemps divergents. Dans les pages qui vont suivre, nous voudrions nous interroger sur la signification que revêt cet intérêt qui confine à l'obsession chez certains auteurs. Nous nous intéresserons plus particulièrement, ce faisant, à sa composante scientifique et technique. Car l'Orient n'est pas perçu uniquement comme l'origine des religions révélées et de l'art. Il apparaît aussi, dans de nombreux écrits, comme le point de départ de l'aventure des sciences et des techniques modernes.

L'ORIENT TERRE NOURRICIÈRE DES SCIENCES ET DES ARTS

L'idée que l'Orient marque le point de départ des sciences et des techniques est assez répandue au XIXᵉ siècle. Saint-Simon et ses disciples, dont il sera fréquemment question ici, offrent l'une des expressions les plus nettes de cette conviction. Saint-Simon s'en fait le premier l'écho dans son *Introduction aux travaux scientifiques du XIXᵉ siècle* de 1808, où l'on peut lire que, du septième au douzième siècle, les Arabes ont été la première nation scientifique du monde[2]. Un peu plus loin, dans cette *Introduction* rédigée au plus fort de sa

---

1. G. Flaubert, *Bouvard et Pécuchet*, Paris, Gallimard, 1979, 412.
2. Cl.-H. de Rouvroy de Saint-Simon, *Introduction aux travaux scientifiques du dix-neuvième siècle*, Paris, 1808, rééd. in *Oeuvres de Claude-Henri de Saint-Simon*, t. VI, Paris, Anthropos, 1966, 9-216, 151 en particulier.

phase " physiciste ", pour reprendre la caractérisation qu'en a donnée Pierre
Ansart[3], Saint-Simon suggère que l'on écrive enfin une histoire des Arabes et
de leur science[4].

Le même credo se retrouve par la suite chez les saint-simoniens qui se réclament de son enseignement. Il figure notamment dans les différents textes consacrés aux sciences par les membres du cercle réuni autour de Prosper
Enfantin[5]. Continuateurs des Grecs et de leur géométrie, inventeurs du zéro et
de l'algèbre moderne, les peuples de l'Orient, les Arabes en particulier, sont à
l'origine des avancées scientifiques de l'Europe.

Il existe bien d'autres témoignages de cette conviction que l'Orient n'est pas
seulement à la source des grandes religions révélées, mais qu'il se situe à l'origine des sciences et des techniques modernes. S'agissant des techniques, on
peut évoquer ici la figure l'architecte Viollet-le-Duc, éminent théoricien de la
construction gothique dans laquelle il voit non pas un style, mais l'essence de
l'esprit bâtisseur moderne, " un moyen de produire bien plus qu'une
production "[6] ainsi qu'il l'explique dans un de ses articles. Pour Viollet-le-Duc
qui se montre d'un nationalisme sans faille, le gothique constitue la manifestation privilégiée du génie de la race française, génie qu'il oppose à celui des
autres nations. Pourtant, à lire plus avant les ouvrages de Viollet-le-Duc, son
Dictionnaire raisonné de l'architecture française et ses *Entretiens sur l'architecture* notamment, on apprend bientôt que le gothique est partiellement d'origine orientale[7]. Il y a quelque chose de troublant dans ce statut qui semble
contredire tout ce que l'architecte écrit par ailleurs sur la spécificité de l'architecture française.

À LA RECHERCHE DU MYTHE

À ce stade, il convient de ne pas oublier tout ce que l'intérêt pour l'Orient
doit au stéréotype d'une origine de la civilisation dans laquelle raison et religion ne seraient pas séparées. Une telle interprétation avait vu le jour au cours
de la seconde moitié du XVIIIe siècle, au travers d'ouvrages comme *Le Monde
primitif* de Court de Gébelin ou *L'Origine de tous les cultes* de Dupuis. Alors

---

3. P. Ansart, *Sociologie de Saint-Simon*, Paris, PUF, 1970.

4. Cl.-H. de Rouvroy de Saint-Simon, *Introduction aux travaux scientifiques du dix-neuvième siècle*, op. cit., 152.

5. Sur la conception saint-simonienne de la science, on pourra consulter A. Picon, " La Science saint-simonienne entre romantisme et technocratisme ", in M. Riot-Sarcey (dir.), *L'Utopie en questions*, Saint-Denis, Presses Universitaires de Vincennes, 2001, 103-123.

6. E.-E. Viollet-le-Duc, *A Monsieur Adolphe Lance, rédacteur du journal* L'encyclopédie d'architecture, extraits de *L'encyclopédie d'architecture* de janvier 1856, col. 10.

7. Concernant cette origine orientale du gothique, voir notamment M. Bressani, *Science, histoire et archéologie. Sources et généalogie de la pensée organiciste de Viollet-le-Duc*, thèse de doctorat dactylographiée, Paris, Université de Paris IV-Sorbonne, 1997.

qu'il se montre si volontiers critique à l'égard de la culture des Lumières, le XIXe siècle recueille cet héritage sans le remettre en cause. Le siècle de l'industrie est obsédé, on le sait, par la divergence entre raison et religion, divergence dont Chateaubriand est un des premiers à s'alarmer dans son *Génie du Christianisme* de 1802. Dans une telle perspective, l'Orient apparaît comme le lieu de l'identité préservée entre raison et inspiration religieuse. Cette identité prend pour la pensée romantique la figure du mythe. On sait à quel point le désir d'enraciner, ou plutôt de réenraciner la raison dans le mythe constitue une préoccupation largement répandue dans les cercles romantiques, même s'il est simultanément question de les distinguer radicalement.

Le visage du XIXe siècle est à cet égard double. D'un côté, l'attitude positiviste qui est la sienne conduit à distinguer les sciences et les techniques de la religion et du mythe. De l'autre, de nombreuses démarches du siècle tendent à rapprocher ces deux séries de termes. Même chez Auguste Comte, s'exprime par moments la tentation d'un rapprochement de ce genre. L'état positif dont il annonce l'avènement s'oppose sans ambiguïté à l'état métaphysique caractéristique de l'ère inaugurée par la Réforme de Luther. Des liens nombreux l'unissent en revanche à l'état théologique des premiers âges de l'humanité, à commencer par l'orientation religieuse qui fait défaut au stade métaphysique de son évolution intellectuelle et sociale[8].

Ce travail simultané de dissociation et de rapprochement fait songer à la façon dont le sociologue des sciences Bruno Latour a pu décrire le régime de la modernité dans son essai *Nous n'avons jamais été modernes*, lorsqu'il écrit que ce régime est marqué par un double travail de purification et d'hybridation, de séparation radicale des sciences et des techniques de toutes les autres activités sociales, en même temps que de perméabilité généralisée entre le scientifique, le technique et ces autres activités[9]. Dans la culture du siècle dernier, la question des rapports entre l'Occident est l'Orient peut s'interpréter comme l'une des formes prises par ce curieux régime de dissociation, qui cache de multiples tentatives de rapprochement. Le XIXe siècle n'est pas aussi " positiviste ", au sens ordinaire du terme, que ce qu'on dit souvent. Le siècle de la machine à vapeur et de l'industrie triomphante est aussi l'un de ceux qui a consacré le plus d'écrits à la religion, sans même parler de son intérêt affiché pour le spiritisme et l'occulte[10].

---

8. *Cf.* A. Petit, *Heurs et malheurs du positivisme. Philosophie des sciences et politique scientifique chez Auguste Comte et ses premiers disciples (1820-1900)*, thèse de doctorat dactylographiée, Paris, Université de Paris I-Sorbonne, 1993.

9. B. Latour, *Nous n'avons jamais été modernes. Essai d'anthropologie symétrique*, Paris, La Découverte, 1997, 20-21 en particulier.

10. Cette ambivalence est très bien rendue dans Ph. Murray, *Le XIXe siècle à travers les âges*, Paris, 1984, rééd. Paris, Gallimard, 1999.

Dans le cas d'une utopie comme le saint-simonisme, ce régime quelque peu contradictoire se trouve ouvertement revendiqué. Tout en insistant sur la supériorité acquise par l'Europe, par la France, l'Angleterre et l'Allemagne en particulier, il s'agit de retremper l'esprit occidental au contact de la matrice orientale dont il est issu. Ce motif revient de manière récurrente dans les écrits rédigés au cours de l'année 1832 par les saint-simoniens en vue de constituer le *Livre nouveau*, sorte d'évangile de la religion qu'ils se proposent de créer[11]. Tel est surtout le sens de l'étonnante aventure des " compagnons de la Femme " qui mène en 1832-1833 les saint-simoniens les plus convaincus à Istanbul puis en Égypte, à la recherche de la femme-messie dont l'union avec Prosper Enfantin doit donner, selon eux, naissance au couple-prêtre de l'avenir[12]. En même temps qu'ils caressent ce projet, on les voit envisager le percement de l'isthme de Suez ou la construction de lignes de chemin de fer. Régénérer la culture européenne, ses sciences et ses techniques, au moyen d'une sorte de retour aux origines, les réarticuler avec la religion et le mythe, telle est l'une des ambitions de leur aventure égyptienne.

Il est à noter que cette idée de ressourcement est générale. Viollet-le-Duc en fait l'un des principes de progrès de l'art et de l'architecture dans un article intitulé " De l'Architecture dans ses rapports avec l'histoire ". " Et notons bien ceci, écrit Viollet-le-Duc, le courant habituel des races supérieures semble se diriger de l'Orient vers l'Occident, mais quand, par exception, un courant contraire s'établit, il en résulte toujours dans l'histoire un développement très puissant, comme si ces peuples indo-européens retrouvaient une nouvelle force vitale en se rapprochant de leur berceau "[13].

SUPÉRIORITÉ OCCIDENTALE ET PROGRÈS DE L'HUMANITÉ

Race supérieure : l'expression fatidique est présente sous la plume de Viollet-le-Duc. Difficile de se voiler la face à ce propos. Le coupe Orient-Occident se révèle inséparable d'un ensemble de stéréotypes qui contribuent à forger une image pittoresque, mais dévaluée, de l'Orient. Comme l'a montré Edward Saïd, La notion d'Orient, est elle-même une construction de la culture européenne destinée à cantonner aux registres du primitif et du pittoresque des peuples dont on présuppose que le développement historique s'est arrêté à un certain stade[14].

11. *Cf.* Ph. Régnier, " Introduction ", *Livre nouveau des saint-simoniens*, Tusson, Du Lérot, 1991, 7-54.

12. Sur l'aventure des compagnons de la Femme, on pourra consulter Ph. Régnier, *Les Saint-simoniens en Egypte (1833-1851)*, Le Caire, Amin F. Abdelnour, 1989.

13. E.-E. Viollet-le-Duc, " De l'Architecture dans ses rapports avec l'histoire ", *Gazette des architectes et du bâtiment*, vol. 4 (1866), 353-364, 361 en particulier.

14. E. Saïd, *L'Orientalisme. L'Orient créé par l'Occident*, 1978, Paris, Le Seuil, 1980 (traduction française).

Les saint-simoniens n'échappent pas à ce cadre idéologique. L'Orient s'assimile pour eux à la femme, à la passivité, au matérialisme, ou encore à la sensualité, tandis que l'Occident incarne l'homme, l'action, la spiritualité et l'ascèse. Dans le même ordre d'idées, l'Orient constitue à leurs yeux une source d'inspiration irremplaçable pour l'art et la littérature, l'Occident étant davantage lié à la science et à ses applications industrielles. Au-delà du saint-simonisme et de sa doctrine, on n'en finirait pas d'énumérer les stéréotypes véhiculés par le couple Orient-Occident au siècle dernier.

De tels stéréotypes viennent à l'appui de l'expansionnisme européen. En dictant sa loi à l'Orient, l'Occident se veut son civilisateur, celui qui le fait revenir dans le cadre de l'histoire et renouer avec la loi du progrès dont il s'était détourné. Au cours de la seconde moitié du XIXe siècle, une réalisation comme le canal de Suez se place clairement sous l'égide de ce retour de l'Orient à l'histoire[15].

Expansionnisme et impérialisme ont partie liée. Dans un essai intitulé *Occident et Orient. Etudes politiques, morales, religieuses* publié en 1835, l'ancien saint-simonien Émile Barrault n'hésite pas à envisager la suzeraineté de la Russie sur l'empire Ottoman, ou encore la tutelle de l'Angleterre sur les pays de la péninsule arabique[16]. Le couple Orient-Occident constituera par la suite la justification idéologique de bien d'autres projets de domination, projets dont certains seront suivis de réalisation. Plus rationnel, capable de séparer la science du mythe et de la religion, même s'il les rapproche en sous-main — ce que Saint-Simon appelle " être organisé " —, l'homme occidental est appelé à dicter, croit-il, sa loi à l'Oriental.

Mais faut-il nécessairement s'arrêter à ce stade ? La justification de l'aventure coloniale n'est peut-être pas seule en cause dans l'affaire. La question des rapports entre l'Orient et l'Occident pourrait bien renvoyer à d'autres enjeux que la mainmise des puissances européennes sur le continent asiatique à des fins économiques et politiques.

TEMPS LINÉAIRE OU ÉTERNEL RETOUR

Revenons pour cela à la citation de *Bouvard et Pécuchet* dont nous étions partis : " Bouvard voit l'avenir de l'humanité en beau. L'homme moderne est en progrès. L'Europe sera régénérée par l'Asie, la loi historique étant que la civilisation aille d'Orient en Occident (...). Les deux humanités enfin seront confondues ". Le plus étonnant, à la lire avec attention, est de constater que

---

15. Voir N. Montel, *Le Chantier du canal de Suez (1859-1869). Une Histoire des pratiques techniques*, Paris, Presses de l'École nationale des Ponts et Chaussées, 1998.
16. É. Barrault, *Occident et Orient. Etudes politiques, morales, religieuses*, Paris, Desessart, A. Pougin, 1835.

bien que l'Europe soit censée être en progrès, c'est elle, et non pas l'Asie, qui se trouve régénérée par la réunion de l'Orient et de l'Occident. Si la loi historique veut que la civilisation aille d'Orient en Occident, quel est donc le statut du retour de l'Occident en Orient, retour régénérateur pour la plus avancée des deux aires culturelles ?

Les saint-simoniens et anciens saint-simoniens constituent là encore des témoins éclairants. Enfantin se montre particulièrement explicite dans les textes qu'il consacre à la question algérienne au début des années 1840. À ses yeux, la colonisation de l'Algérie raterait son but si elle ne constituait pas pour les Français une occasion de progresser : " Notre occupation d'Afrique n'aurait pour ainsi dire pas de sens, ou plutôt serait une vraie niaiserie, si elle n'était que ce que notre orgueil prétend qu'elle est, c'est à dire un moyen de civilisation pour les Arabes. Elle est avant tout un moyen de civilisation pour les Français[17] ". Qu'a donc à apprendre l'Occidental de cet Oriental qu'il se propose d'initier aux complexités de la société industrielle ?

Pour les saint-simoniens, la réponse est assez claire. Tout à leur désir de réconcilier raison et religion, spiritualité et matérialisme, ils se montrent fascinés par les sociétés orientales auxquelles ils attribuent des vertus à la fois religieuses, matérialistes et sensuelles. Le caractère théocratique qu'ils prêtent à ces sociétés leur paraît en particulier digne d'être imité. Dans ses textes sur l'Algérie, Enfantin ajoute à cela le caractère collectif de la propriété rurale[18]. Selon lui, ce caractère collectif constitue un meilleur point de départ que l'individualisme économique forcené des européens pour instaurer l'" association universelle " des hommes. Pour reprendre l'opposition à laquelle recourra plus tard Ferdinand Tönnies[19], l'Orient permet de retourner de la *Gesellschaft* à la *Gemeinschaft*, de la société à la communauté, ou plutôt de les concilier, ce qui semble à première vue impensable. Comment rendre son âme collective à une société composée d'individus ? Il s'agit là d'un des problèmes essentiels auxquels se trouve confrontée la pensée du XIXe siècle. L'importance accordée à la dimension religieuse participe du désir de restaurer une véritable communauté par l'intermédiaire d'une foi partagée.

Les sciences et les techniques ne sauraient rester à l'écart de cette interrogation lancinante. Peuvent-elles continuer à progresser au sein d'une société divisée en partis et en classes antagonistes ? Il n'est pas fortuit que les saint-simoniens mettent l'accent sur la dimension nécessairement collective du pro-

17. P. Enfantin, lettre à Arlès-Dufour du 17 juin 1840, Bibliothèque de l'Arsenal Fonds Enfantin Ms 7839.

18. Voir notamment P. Enfantin, *Colonisation de l'Algérie*, Paris, P. Bertrand, 1843.

19. F. Tönnies, *Gemeinschaft und Gesellschaft*, 1887, New York, Harper, 1963 (traduction anglaise). Sur l'opposition entre communauté et société, voir par ailleurs R.A. Nisbet, *La Tradition sociologique*, New York, 1966, Paris, PUF, 1984 (traduction française).

grès scientifique et technique en même temps qu'ils se préoccupent de réunir l'Orient et l'Occident. C'est en Orient qu'ils vont tenter de constituer le noyau de leur " armée pacifique des travailleurs " après s'être heurté à d'insurmontables obstacles en France.

Mais en filigrane de ces justifications du ressourcement oriental se dessine peut-être un motif encore plus fondamental, celui-là même que révèle involontairement Viollet-le-Duc dans son article " De l'Architecture dans ses rapports avec l'histoire ", en donnant à ce ressourcement des allures de retour en arrière, d'inversion de la flèche du temps, de restauration du passé. Le thème de la restauration du passé est également très présent sous la plume des saint-simoniens. Il affleure dans de nombreux passages du célèbre *Système de la Méditerranée* de Michel Chevalier : " Qu'elles étaient brillantes les galères qui, portant la fleur des chevaliers de l'Occident, allèrent, après avoir en passant soumis Zara, asseoir Baudouin de Flandre sur le trône de Constantin et inaugurer le lion de Saint-Marc en Morée et dans les îles de l'archipel ! Eh bien ! Venise lancera de son sein de nouveaux convois plus magnifiques "[20]. " Oui, nous devons faire renaître de ses cendres la bibliothèque d'Alexandrie, ressusciter les grandes momies de Memphis que nous avons dépouillées de leurs bandelettes pourries et aider le Christ à retrouver sa tombe et son berceau "[21], s'enthousiasme quant à lui Enfantin.

Au thème de la fusion avec l'Orient se surimpose celui de la réconciliation entre le passé et le présent de l'Occident. Dans cette perspective, l'Orient en vient à incarner des époques et des traits de la culture occidentale qu'on aurait pu croire révolus mais qui sont susceptibles de revenir au premier plan par l'intermédiaire d'une sorte de court-circuit temporel. Coloniser l'Algérie, c'est revenir au Moyen Age collectif et théocratique à l'heure des machines à vapeur et des premiers chemins de fer. Le contact avec l'Orient provoque d'autres retours, sur des échelles plus brèves de temps. Lorsque le saint-simonien Charles Lambert fonde par exemple l'École polytechnique du Caire au milieu des années 1830, il cherche à renouer avec le projet pédagogique qu'avait conçu Gaspard Monge sous la Révolution, au lieu de s'inspirer de la formation dispensée à son époque par une École Polytechnique qui s'est détournée de l'esprit de ses fondateurs[22].

---

20. M. Chevalier, *Religion saint-simonienne. Politique économique. Système de la Méditéran-née*, Paris, bureaux du *Globe*, 43.

21. P. Enfantin, lettre à Arlès-Dufour du 18 janvier 1840, Bibliothèque de l'Arsenal Fonds Enfantin Ms 7839.

22. Sur la création de l'école d'ingénieurs du Caire, on pourra consulter G. Alleaume, *L'École polytechnique du Caire et ses élèves. La Formation d'une élite technique dans l'Egypte du XIXe siècle*, thèse de doctorat dactylographiée, Lyon, Université Lyon II, 1993. En ce qui concerne l'École polytechnique et son évolution au cours de la première moitié du XIXe siècle, voir B. Belhoste, A. Dahan-Dalmédico, A. Picon (dir.), *La Formation polytechnicienne 1794-1994*, Paris, Dunod, 1994.

" Ils reviendront ces dieux que tu pleures toujours! Le temps va ramener l'ordre des anciens jours. La terre a tressailli d'un souffle prophétique "[23], écrit Gérard de Nerval dans *Les Chimères*. Dans la littérature romantique de la première moitié du XIXᵉ siècle, l'Orient se voit fréquemment associé aux thèmes du temps cyclique et de l'éternel retour. Cette association perdurera par la suite dans de nombreux poèmes et romans.

La persistance de ces thèmes renvoie probablement à la complexité des perceptions du temps au sein de la société industrielle naissante, une société qui se souvient, surtout en France, de ses origines agricoles et de l'importance de la succession des saisons, une société qui hésite sur le sens à donner au mot révolution. Si le mot renvoie pour les uns, qu'ils soient républicains ou bonapartistes, à l'idée d'un changement irréversible, il se comprend pour les autres, pour les ultras souvent marqués par la lecture de Joseph de Maistre, en particulier, dans son acception ancienne de cycle astronomique ou historique. Dans cette acception ancienne, les choses reviennent à leur point de départ, les astres repassent au mêmes endroits du ciel, les rois sont restaurés et l'ordre rétabli[24].

Le temps linéaire du progrès scientifique et technique semble à première vue incompatible avec la pluralité des perceptions du temps social et historique qui caractérise le XIXᵉ siècle. La tension qui s'établit du même coup pourrait devenir insupportable sans l'existence de mécanismes de compensation idéologiques et symboliques. La perspective d'une régénération de l'Occident au moyen de l'Orient fait probablement partie de ces mécanismes de compensation permettant de réintroduire la perspective de retours en arrière et de cycles au sein de la marche en avant des sciences et des techniques. Au travers de la fusion annoncée de l'Occident et de l'Orient, la modernité rencontre l'immémorial, le progrès s'allie avec son contraire.

Au cours de la première moitié du XIXᵉ siècle, une telle alliance trouve son illustration la plus frappante avec la recréation d'un État grec. Plus encore que la politique de modernisation menée par Méhémet Ali en Égypte, cette recréation permet de conjuguer progrès et récit des origines, temps linéaire et temps cyclique, Orient mythique et Occident parti à la recherche de son passé. " Ils reviendront ces dieux que tu pleures toujours ! Le temps va ramener l'ordre des anciens jours. La terre a tressailli d'un souffle prophétique ".

---

23. G. de Nerval, " Myrtho ", 1854, in *Œuvres*, t. 1, Paris, Gallimard, 1974-1978, 12.

24. Sur ce conflit entre conceptions de l'événement révolutionnaire, on pourra consulter M. Ozouf, *Les Aveux du roman. Le Dix-neuvième siècle entre Ancien Régime et révolution*, Paris, Fayard, 2001.

FAIRE L'HISTOIRE DES USAGES DES OBJETS TECHNIQUES : FORMULES, PROJETS ET PRATIQUES. L'EXEMPLE DES BROUETTES SUR LE CHANTIER DU CANAL DE SUEZ (1859-1869)

Nathalie MONTEL

En 1980, paraissait dans la revue *La Recherche*, revue française de vulgarisation scientifique, un article signé Bertrand Gille, intitulé " Petites questions et grands problèmes : la brouette "[1]. Dans cet article, l'historien des techniques s'interrogeait sur l'étymologie du mot " brouette ", mais aussi sur les origines et l'époque d'apparition en France de cet instrument, devenu depuis un objet familier et d'usage commun, que l'on rencontre aujourd'hui fréquemment dans nos campagnes et jardins citadins. C'est ce même instrument, d'apparence anodine, qui est également au centre de l'enquête menée ici. La perspective est néanmoins différente. Il ne s'agira pas de savoir si la brouette est vraiment née en Chine ou si elle est effectivement apparue en France dans la première moitié du XIIIe siècle mais d'illustrer, à travers l'histoire de ses usages au XIXe siècle, la dimension culturelle que portent en eux les objets techniques.

Il n'est pas question de faire un point complet sur les usages de la brouette à travers le monde, mais plus simplement d'évoquer quelques-uns des moments de cette histoire en prenant notamment appui sur une recherche consacrée à l'histoire du chantier du canal de Suez[2], qui fut l'occasion de croiser cet outil. Avant de faire état des péripéties de la brouette au cours de ces travaux en Égypte, il est toutefois nécessaire de procéder à un rapide état des savoirs relatifs à l'emploi de cet outil dans les travaux publics. C'est sur le terrain de la presse spécialisée que se terminera l'enquête. De la confrontation des résultats issus de ces différents terrains d'investigation, quelques enseignements de nature méthodologique pourront être tirés.

---

1. Cet article a été repris dans Coll., *La Recherche en histoire des sciences*, Paris, Seuil, 1983, 79-88 (Point Sciences).
2. Voir Nathalie Montel, *Le chantier du canal de Suez (1859-1869). Une histoire des pratiques techniques*, Paris, Presses de l'École nationale des Ponts et Chaussées/Éditions In Forma, 1998.

LA MISE EN FORMULE DE L'USAGE DE LA BROUETTE

Les hommes qui débarquent en Égypte en 1859 pour prendre en charge cette entreprise colossale qu'est le percement de l'isthme de Suez arrivent pour la plupart de France. Ils traversent la Méditerranée avec, dans leur bagage, les théories et les savoir-faire techniques en vigueur dans leur pays. C'est sur le contenu de ce bagage qu'il convient, dans un premier temps, de s'arrêter brièvement, en concentrant l'attention sur la principale tâche qui attend les responsables techniques de ces travaux, c'est-à-dire les terrassements.

Dans ce domaine, les références les plus anciennes en matière de savoirs et de savoir-faire remontent au XVIIᵉ siècle et se trouvent chez les militaires auxquels on confie les travaux de fortification. La réalisation des places fortes du royaume nécessite en effet d'importants mouvements de terre et la brouette joue dès cette époque un rôle central sur les chantiers de terrassement. L'un des premiers à s'intéresser de très près aux détails des travaux d'un point de vue non plus simplement technique mais économique est Vauban[3]. Ce qui motive notamment cet ingénieur royal devenu commissaire général des fortifications (1678), c'est la volonté de réduire les dépenses de ces chantiers fort coûteux et de fixer des prix qui réalisent un compromis entre les intérêts du roi, ceux des entrepreneurs et ceux des soldats employés comme terrassiers. Dans une instruction datée de 1688, il préconise de substituer à l'arbitraire, qui existait auparavant dans le mode de fixation des prix, une méthode consistant à rémunérer la toise cube de terre chargée dans une brouette puis roulée sur une certaine distance, en fonction du salaire journalier que l'on souhaite faire gagner au terrassier[4]. Le détail de la méthode imaginée par Vauban repose sur des calculs faisant intervenir des données moyennes issues de ses propres expériences. Il estime ainsi que deux toises cubes de terre, qui est la tâche journalière commune assignée à un ouvrier de force moyenne, peut être menée en cinq cents brouettées à une distance de quinze toises. La brouette, qui est l'instrument permettant de donner la mesure des volumes déblayés puis transportés par les terrassiers, se trouve ainsi au centre des estimations et participe à l'évaluation des différentes données expérimentales prises en compte dans ces calculs.

Au siècle suivant, les officiers du génie reprennent les principes de la méthode imaginée par Vauban et les chantiers de terrassement deviennent le

---

3. Voir notamment Anne Blanchard, *Vauban*, Paris, Fayard, 1996.

4. Vincennes. Service historique de l'armée de terre (SHAT). IV, article 21, section 6. Vauban, *copie du règlement fait en Alsace pour le prix que les entrepreneurs devaient payer aux soldats employés au transport et remuement des terres de la fortification des places de sa Majesté*, 1688. La toise vaut approximativement deux mètres.

théâtre d'observations et d'expériences de plus en plus nombreuses[5]. Ces expériences visent à mesurer de façon toujours plus précise la capacité des hommes au travail, les volumes de terre déblayés, les distances sur lesquelles elles peuvent être transportées mais aussi le temps passé à accomplir les différentes opérations qu'une analyse de plus en plus fine du travail met en évidence. Les réflexions issues de ces expériences mettent progressivement l'accent sur les pertes de temps observés et sur la nécessaire coordination qui doit exister entre les deux principales activités que sont " la fouille " et " le roulage ". Elles conduisent à imaginer une organisation idéale des ateliers de terrassement, fondée sur un maniement optimisé de la brouette et dans laquelle les différentes manœuvres s'enchaîneraient en un mouvement continu.

Un tournant s'opère au début du XIXe siècle. Commence en effet une nouvelle période, marquée par la recherche d'une loi générale et ponctuée des multiples tentatives en vue d'élaborer une formule universelle. Dès la fin des années 1810, les premières théories sont échafaudées et des équations apparaissent, qui tiennent compte des notions récemment élaborées par la physique, comme celle de travail[6]. Réaliser des terrassements à la brouette revient dès lors à résoudre un problème mathématique, que le lieutenant-colonel du Génie Picot, dans un mémoire daté de 1834, formule de la manière suivante : " L'étude du terrassement à la brouette conduit à l'examen de cette question. Quelles sont les quantités de travail du chargeur et des rouleurs pour des terrassiers de force moyenne travaillant à la tâche, et comment faut-il répartir les ouvriers d'un atelier pour que le travail définitif soit un maximum, ou que le prix du terrassement soit un minimum "[7].

Dans les aide-mémoire et manuels à l'intention des ingénieurs, les collections de données issues de l'expérience cèdent désormais la place à des équations incluant les différents paramètres à prendre en compte, comme la nature du sol, le prix à payer aux terrassiers, ou la distance que doivent parcourir les rouleurs avec leur brouette. Ces formules et analyses modèles du travail sur les chantiers fournissent aux ingénieurs d'État des références utiles à la rédaction

5. Vincennes. SHAT. IV, article 21, section 6. Parmi les nombreux mémoires conservés : *Instruction pour le deblay et transport des terres ou lon rappelle celle de M. le maréchal de Vauban fait en 1688, 1749 ; Filley, Observations sur la manière la plus exacte et la plus équitable que l'on doit suivre pour fixer le prix des terres à payer aux soldats travaillant à l'attelier*, mémoire adressé à la cour en 1756 et envoyé au Comte de Saint-Germain en avril 1776 ; Senermont, *Observations sur le transport et le remuement des terres*, 14 janvier 1796.

6. Vincennes. SHAT. IV, article 21, section 6. Notamment : *Notice sur les déblais*, par M.A. Vène, capitaine du Génie, 1819 ; *Quelques observations sur les déblais*, par M. Villeneuve, capitaine du Génie à Toulon, avril 1823 ; *Des terrassements*, par le chef de bataillon du Génie, ingénieur en chef Répécaud, décembre 1822.

7. Vincennes. SHAT. IV, article 21, section 6. Picot, *Mémoire sur les terrassements à la brouette*, 1834.

des devis et leur donnent des repères à la fois pour évaluer les coûts des travaux et négocier les prix avec les entrepreneurs qui emploient les terrassiers.

C'est nantis de ces savoirs, mais aussi de leurs expériences des chantiers en Europe, que les techniciens français recrutés par la Compagnie universelle du canal maritime de Suez arrivent dans l'isthme égyptien à partir de 1859.

LA BROUETTE : PIERRE D'ACHOPPEMENT D'UNE TENTATIVE DE TRANSFERT TECHNIQUE EN ÉGYPTE

Le volume total des terrassements à réaliser sur ce chantier est sans précédent et les premières estimations prévoient qu'il sera nécessaire d'employer huit mille ouvriers pendant cinq ans, en évaluant le rendement de chacun à un mètre cube et demi en moyenne par jour, pour enlever à bras d'homme la partie des terres situées au-dessus du niveau de la mer[8]. La période des premières installations passée, il s'agit d'examiner de plus près les procédés à mettre en œuvre pour organiser au mieux ces travaux et surtout tirer le meilleur parti de la main d'œuvre mise à la disposition de la Compagnie par le gouvernement égyptien. S'inspirant des pratiques qui ont cours en France, les ingénieurs de ce chantier projettent alors de recourir à des procédés faisant intervenir les outils usuels : des brouettes, des pelles, des madriers, mais aussi d'appliquer sur certains tronçons à creuser des dispositifs un peu plus sophistiqués, combinant ces mêmes outils avec des câbles et des poulies. Ces dispositifs de conception nouvelle sont baptisés " brouettes volantes " et " brouette à la corde "[9].

En étudiant les pratiques techniques de ce chantier, à partir des multiples traces laissées par l'activité des agents de la Compagnie de Suez, on s'aperçoit que les dispositifs ainsi imaginés n'ont en fait pratiquement pas fonctionné, que très peu de mètres cubes de terrassement ont en définitive été réalisés au moyen de ces différents procédés. Ces engins se sont révélés peu opérationnels dans ce désert de sable qu'est l'isthme de Suez. Ils ont dû être rapidement abandonnés. À l'origine de cet abandon, on ne trouve pas seulement des procédés probablement trop compliqués à mettre en œuvre sur une grande échelle, mais une raison plus profonde que les agents de la Compagnie découvrent avec surprise et à leurs dépens, à savoir que la brouette, qui est d'un usage banalisé en Europe dans les travaux de l'agriculture et du génie civil, est inconnue en Égypte[10]. De surcroît, les diverses tentatives effectuées en vue de faire utiliser

---

8. Linant-Bey et Mougel-Bey, " Avant-projet du percement de l'isthme par un canal maritime entre Péluse et Suez, 20 mars 1855 ", in Ferdinand de Lesseps, *Percement de l'isthme de Suez. Exposé et documents officiels*, Paris, Plon, 1855, 126.

9. Voisin-Bey, *Le canal de Suez. Atlas*, Paris, H. Dunod et E. Pinat, 1906, Planche n° XXXVI.

10. Voir Nathalie Montel, " L'organisation du travail sur les chantiers de terrassement ", *Culture technique*, n° 26 (1992), 135-145.

la brouette par les fellahs égyptiens se soldent toutes par des échecs. Un témoin présent lors de l'une de ces tentatives rapporte la scène suivante : " J'ai vu, de mes yeux vu, la solution suivante adoptée par trois manœuvres arabes, après maintes tentatives infructueuses pour suivre les leçons du chef d'équipe européen. L'un prenait la roue ; les deux autres les brancards de la brouette remplie, et mes trois gaillards de porter triomphalement cette charge jusqu'au point fixé pour jeter le déblai ; arrivés là, retournement de la brouette, puis retour au point de départ "[11].

Cette anecdote rend compte de la difficulté d'apprentissage de l'usage de la brouette, mais elle illustre aussi un fait plus général, à savoir que l'objet technique ne porte pas en lui l'usage qui en est fait, encore moins les gestes qui lui sont associés.

Lorsque l'on mène l'enquête de manière plus large, on s'aperçoit que les hommes de la Compagnie de Suez ne sont ni les seuls ni les premiers à avoir rencontré de tels déboires ou à faire le constat de la méconnaissance de l'outil. Parmi les exemples qui en témoignent[12], l'un se situe en 1798, à l'époque où l'expédition de Bonaparte amène au Caire de nombreux Français. Un notable de la ville a consigné dans son journal ses observations sur les divers événements qui l'ont marqué durant cette période de la présence française. Entre autres faits rapportés, on peut relever cette description : " Ils [les Français] recouraient à des instruments faciles à manier et épargnant la peine, ce qui permettait une exécution rapide des travaux. Ainsi, au lieu de paniers ou de récipients, ils utilisaient de petites charrettes qui avaient deux bras allongés par derrière ; on les remplissait de terre, d'argile ou de pierres [...] ensuite on prenait en main les deux bras, on poussait devant soi et la charrette roulait sur sa roue avec la moindre peine jusqu'au chantier ; on les vidait enfin, en la penchant d'une main, sans aucune fatigue "[13].

Ce témoignage rend compte d'une vision certes subjective, et sans doute idéalisée à sa manière, de l'usage de la brouette, mais il renseigne néanmoins sur la curiosité que suscite l'emploi de l'instrument et surtout atteste de la méconnaissance qu'on en avait dans le pays. La situation semble sur cette question avoir peu évolué, puisque les archéologues de l'Institut français

11. Olivier Ritt, *Histoire de l'isthme de Suez*, Paris, Hachette, 1869, 181-182.

12. Voir par exemple le témoignage de l'ingénieur des Ponts et Chaussées Stoecklin, chargé pour le compte du gouvernement égyptien, de surveiller les travaux de construction du bassin de radoub du port de Suez : *Notice sur le bassin de radoub de Suez*, Bordeaux, Imprimerie A. Bord, 1867, 64. On peut également, dans le cas de la Turquie, se reporter à l'expérience d'un officier du génie français qui relate la résistance opposée à l'utilisation de la brouette. Cette expérience est rapportée par Frédéric Hitzel : " La France et la modernisation de l'Empire ottoman à la fin du XVIIIe siècle ", in Patrice Bret (ed.), *L'expédition d'Égypte, une entreprise des Lumières 1798-1901*, Cachan, Technique & documentation, 1999, 12.

13. Abd al-Rahmân al-Jabartî, *Journal d'un notable du Caire durant l'expédition française, 1798-1801*, traduit et annoté par Joseph Cuoq, Paris, Albin Michel, 1979, 89.

d'archéologie orientale du Caire se heurtent aujourd'hui au même type de difficulté lorsqu'ils prescrivent l'utilisation de brouettes sur leurs chantiers de fouille.

Les rapports de chantier des travaux du canal de Suez révèlent que les ouvriers égyptiens en sont finalement revenus à leurs anciennes méthodes : la plupart des terrassements manuels sont réalisés par des techniques locales ancestrales et au moyen d'outils qui leur sont familiers, c'est-à-dire une petite pioche permettant d'attaquer la terre, et des couffins ou paniers tressés en tiges de palmier, pour la transporter[14]. Entre cinq et dix millions de mètres cubes de terre et de sable sont extraits de cette façon, ce qui représente environ un dixième du volume total de terre que nécessitait le creusement du canal de Suez. Cette difficulté d'acclimatation de la brouette en Égypte suggère l'ampleur des difficultés inhérentes à la réalisation de travaux à l'étranger et donne à voir certains des obstacles culturels auxquels peuvent se heurter des tentatives de transfert de technique.

Pour tenter de mieux comprendre cette difficulté d'acclimatation, l'ethnologue André Georges Haudricourt fournit une piste intéressante. Au détour d'un courrier, il écrit : " Le problème de la brouette est intéressant, elle est venue chez nous par le Nord [...]. Le problème est de voir pourquoi elle ne s'est pas rapidement répandue vers le sud. C'est que la brouette est un véhicule à moteur humain, son usage fait partie de ce que M. Mauss appelle les techniques du corps c'est-à-dire des habitudes musculaires. Les habitudes musculaires et les habitudes mentales sont les plus tenaces. Elles ne peuvent se modifier qu'à la faveur d'une modification d'ensemble "[15].

Ainsi l'échec de la tentative de transfert de cette technique de transport des terres sur le sol égyptien s'explique en grande partie par l'impasse faite sur la dimension culturelle de l'outil préconisé, un outil qui met en jeu le corps humain et requiert un certain type d'effort physique auquel le paysan égyptien envoyé dans l'isthme de Suez pour participer aux travaux n'est pas habitué.

DE LA DIFFUSION DES DESSINS DE BROUETTES AUX ENSEIGNEMENTS POUR L'HISTORIEN DES TECHNIQUES

Pour autant, la carrière des dispositifs incluant des brouettes imaginés pour ce chantier ne s'arrête pas là. Ils connaissent en effet un certain succès éditorial. À l'époque du chantier, les procédés conçus pour effectuer les terrasse-

---

14. Paris, Archives nationales, 153 AQ/1610A : rapport mensuel du directeur général des travaux, octobre 1861.

15. André-Georges Haudricourt, *La technologie science humaine - Recherches d'histoire et d'ethnologie des techniques*, Paris, Éditions de la Maison des Sciences de l'Homme, 1987, 310-311, extrait d'une lettre à Charles Parrain du 22 octobre 1936.

ments font l'objet de représentations graphiques. Ces dessins, sur lesquels de petits personnages en action ont été figurés pour bien en faire comprendre le fonctionnement, sont ensuite reproduits et publiés[16]. Dans l'atlas relatif à l'histoire des travaux, qu'il fait paraître en 1906, l'ancien directeur du chantier choisit de consacrer l'une des quarante planches à ces " modes d'exécution primitivement projetés pour l'ouverture de la rigole maritime de la traversée du seuil d'El-Guisr ", montrant ainsi son attachement, comme sa probable implication personnelle dans la conception et le choix de ces dispositifs[17]. Mais bien avant que Voisin ne fasse connaître ces dessins, ils avaient déjà paru dans plusieurs périodiques destinés aux professionnels du génie civil. Ils se trouvent notamment dès 1862 dans la revue *Les nouvelles annales de la construction*[18], puis dans un ouvrage publié trente ans plus tard intitulé *Procédés et matériaux de construction*[19]. Celui-ci est signé par Alphonse Debauve, ingénieur des Ponts et Chaussées et auteur prolixe de manuels sur les techniques de travaux publics. Dans ces deux cas, les dessins sont commentés et présentés comme étant les procédés mis en œuvre pour réaliser les terrassements manuels du chantier du canal de Suez. La séduction exercée tant par l'ingéniosité que par la nouveauté de ces dispositifs l'a vraisemblablement emporté sur la nécessité de vérifier leur emploi effectif.

Grâce à la brouette, qui a servi de fil conducteur, des formules, des projets et des pratiques de chantier ont successivement été évoqués. On a vu comment cet outil élémentaire avait pu être au cœur d'une conception rationalisée de l'organisation des chantiers de terrassement, comment elle se révéla être la pierre d'achoppement du transfert en Égypte de pratiques techniques en vigueur en France, comment enfin des informations erronées relatives à son usage avaient circulé.

Au-delà du cas particulier, il y a là matière à tirer quelques enseignements d'ordre plus général. Ces péripéties de la brouette invitent en effet l'historien des techniques à adopter une attitude prudente et à s'interroger sur l'utilisation qu'il fait des sources à sa disposition. Selon les types de sources que l'on mobilise, c'est en effet des registres distincts de l'histoire des techniques que l'on saisit et auxquels on accède. Ces registres sont multiples : ils vont de l'histoire des usages et des pratiques à l'histoire des représentations de la technique, en passant par l'histoire de la conception et de l'ingéniosité, pour ne citer

---

16. Ils paraissent pour la première fois dans le journal créé par la Compagnie universelle du canal maritime de Suez et destiné notamment à ses actionnaires : *Journal de l'isthme de Suez*, n°114 (mars 1861), 89.

17. Voisin-Bey, *Le canal de Suez. Atlas*, *op. cit.*

18. A. Cassagnes, " Percement de l'Isthme de Suez ", *Nouvelles annales de la construction* (janvier 1862), col. 6-12.

19. Alphonse Debauve, *Procédés et matériaux de construction*, tome 1 : *Sondages ; terrassements ; dragages*, Paris, Vve Ch. Dunod, 1884, 200-202, réédité en 1894.

que quelques exemples. Ces différents registres ne sont évidemment ni exclu-
sifs ni indépendants l'un de l'autre, mais au contraire étroitement mêlés. Toutefois, il convient de noter que chacun d'eux possède une trajectoire
particulière et des temporalités qui lui sont propres. Si les différentes histoires
produites peuvent se rencontrer en certains lieux ou converger à certaines époques, il est important, à mon sens, de ne pas oublier qu'elles ne coïncident pas
forcément, de ne pas les confondre ou tenter d'impossibles déductions de l'une
à l'autre. Le cas des brouettes illustre les discordances et les écarts qui peuvent
exister entre des techniques modélisées, telles qu'elles apparaissent au travers
de mémoires théoriques ou de devis et la manière dont elles sont mises en
œuvre sur le terrain, mais aussi entre la conception d'un travail et la façon dont
il est finalement réalisé, ou encore entre les techniques décrites par la littérature spécialisée ou les manuels d'enseignement et celles couramment
employées. Pour qui cherche à saisir les usages passés effectifs d'objets techniques[20], et s'interroge en particulier sur leur composante culturelle, le choix
des sources apparaît donc déterminant.

---

20. Comme l'y invite notamment David Edgerton dans son article : " De l'innovation aux usages. Dix thèses éclectiques sur l'histoire des techniques ", *Annales Histoire, Sciences Sociales*
(Juillet - Octobre 1998), n° 4-5, 815-837.

ATTITUDES, ACTIVITIES AND ACHIEVEMENTS :
SCIENCE IN THE MODERN MIDDLE EAST

Yakov M. RABKIN

INTRODUCTION

Western influence used to be commonly seen as the exclusive factor of scientific developments in the region. Middle East was presented as virgin soil on which no native plants of scientific creativity were to be found and only memories of the Islamic Golden Age redeemed the region from total obscurantism. In the last decades of the 20th century, this view came under review, and elements of continuity between Medieval Arabic science and modern scientific research were gradually brought to light[1]. The new emphasis on continuity and native origins of modern science enhanced the acculturation of modern science in the Middle East undergoing resurgence of Islam and Judaism in the 1980s and 1990s. This view helped reduce the traditionalists' mistrust of contemporary science and, particularly, of its modernizing purveyors among local elites.

Sciences were shown to be part of traditional Islamic scholarship. Mathematics continued to be developed in Arabic, the classical *lingua franca* of science, not only in Arabic-speaking lands such as Egypt and Syria but in Iran, India and Turkey well into the 19th century. Research was conducted, scientific books were published, and the infrastructure was in place to receive and integrate Western science within the local cognitive traditions. However, rather than rely on local scientific efforts, Western scientific expansion obliterated or further marginalized Islamic science. It emphasized a conflict between science and tradition in spite of the fact that science was an integral part of Islamic and Judaic scholarship well into the 20th century. For example, scientific education of the prominent Judaica scholar Rabbi Yosef Kafah (1917-2000) included studies of Ptolemaic astronomy from manuscripts with a Muslim in his native Yemen.

---

1. See, e.g., Ekmeleddin Ihsanoglu (ed.), *Transfer of Modern Science and Technology to the Muslim World*, Istanbul, IRCICA, 1992.

CHANNELS OF TRANSFER

Modern science entered the Middle East through two main channels : new educational institutions and publication of translations from Western languages. The Ottoman Empire developed contacts with European science as early as the 15th century : the influx of Jewish doctors expelled from Spain vastly improved the state of medicine while the annexation of Bosnia and Serbia contributed to the development of military industries. Translations of scientific texts as well as reports filed by Ottoman envoys from Europe kept Istanbul elites abreast of Western scientific developments. Ottoman ambassadors also paid visits to the Imperial Academy of Sciences in St Petersburg, an impressive case of the diffusion of Western science into a hitherto scientifically deserted land. Russia, an aggressive military foe, fascinated the Ottomans, and particularly the Turkish nationalists in the 20th century, who would maintain a keen interest in Russia's attempts to cultivate science.

The State was behind most of scientific developments in the Middle East. They began with the establishment of military schools in Istanbul, Cairo and Tehran in the 18th and 19th centuries. European expatriates mostly taught technological subjects while the sciences were initially taught by traditional Islamic scholars. Egyptian scholars remained unimpressed by chemical experiments staged by the occupying French, but Muhammed Ali, who attempted to modernize Egypt in the first half of the 19th century, directly attributed the success of the French to their superior science. He adopted the French style of secular education, promoted daily press, and encouraged the translation of important works into Arabic. Most of these activities were undertaken for the benefit of the army, all of them — for the benefit and under the tutelage of the State.

Military considerations were essential motive forces behind the transfer of Western science into Iran and the Ottoman Empire. Conversely, the early Zionist settlers at the turn of the century frowned upon science and relied on imported rifles for their elementary military needs. Israel's Jewish population would come to rely on science for security purposes only after the declaration of statehood in 1948 ; but by the end of the century it would become a major producer and exporter of high-tech military wares.

Both military technical schools and a few civilian institutions which had been established by mid-19th century facilitated the assimilation of Western technologies and artifacts. They also produced a literate segment of teachers and government officials who challenged traditional ideas of education, religion and social reform. The first civilian institutions of higher learning produced locally trained administrators, financiers, engineers, and medical doctors. Some, like the *Dar al-Funun* in Tehran, housed the government print-ing press and disseminated modern cosmopolitan ideas in society at large.

The earliest appearance of modern civilian science in the region can be attributed to Arabic-speaking Christians and Muslims in the second half of the 19th century. Christians of all sects and Jews had their own school systems, and operated with a relatively free hand. Several universities and colleges in the region trace their origins to foreign missionary activities begun in the 19th cen-tury. The American Universities in Beirut and Cairo, the Robert College, later renamed the Boğaziçi University, in Istanbul, the Alborz College of Tehran, all of them played important roles as conduits of Western attitudes that affected generations of local elites. Most of these missionary activities originated in an Anglo-Saxon revivalist movement that was a religious response to the indus-trial revolution and the social problems it had engendered. Modern science was no more than a by-product of the missionary efforts in the Middle East.

The French influence was also felt in the region. The Saint-Simonians, a humanistic movement also of French origin, spread their altruistic work in medicine and science in the 1830s. A few decades later a *lycée français* opened its doors on the Bosphorus, while in Beirut the Jesuits established Collège St Joseph later in the century. The *Alliance israelite universelle* which opened schools throughout the Middle East, also introduced French influence and modern education, mostly, but not exclusively, among the Jews. The use of English and French as respective languages of instruction brought the elites closer to modern science but, on the other hand, constituted an obstacle to broader dissemination of scientific knowledge.

Muhammed Ali in Cairo as well as his counterparts in Istanbul and Tehran sent students to Europe as a means to transfer science and technology expertise into their respective countries. The expatriation of thousands of students to European and American universities, particularly for graduate studies, became the standard mode of scientific training in the 20th century. While many of these expatriates failed to return to their native lands, the impact of Western education would have a long lasting effect on the development of science in the Middle East.

These educational innovations were controversial. Even avowed moderniz-ers observed that " the new education was foreign, odd and unpopular "[2]. It came to be divorced from the traditional system of values and the resulting duality hampered modernization efforts, insofar as it relied on " a science con-

---

2. Yusef Salah El-Din Kotb, *Science and Science Education in Egyptian Society*, New York, Teachers College, Columbia University, 1951 (Issue 967).

ceived and produced by foreign scientists and technologists "[3]. This practice proved to be remarkably durable throughout the Middle East. The tradition of transfer of technology through turnkey projects permeated the history of the 19th and 20th centuries in most of the region. It was only in the 1970s that Turkish and Israeli companies emerged as important regional purveyors of such technologies.

University-type institutions of local provenance would appear in the Middle East long after the military schools. In the wake of Russia which had fomented research by expatriate scientists on its soil before training native scientists, the Sublime Porte also began by opening an academy of sciences, *Encümen-i Daniş*, in 1850, long before the first university would be born in the Ottoman realm. Unlike the Russian Academy, the *Encümen-i Daniş* was to follow intellectual, including scientific, developments in Europe rather than undertake original research on its premises. This showed an appreciation of science, but science as a product rather than a process.

After several aborted attempts to open a university, it was at the turn of the century that the Imperial University, *Darülfünun-ı Şahane*, was started in Istanbul. It contained the departments of sciences and mathematics alongside those of medicine, law and religious studies. The university was later modernized with the help of German expatriates and opened its doors to first female students. The integration of women into the sciences and the professions became a distinctive trait of the late Ottoman, and particularly of the Turkish republican pattern of education. Modern universities were established in Istanbul (1944) and Ankara (1946). By 1967 there were eight universities, and by the end of the 20th century a network comprising a threescore of universities covered practically every large city in the republic.

Ottoman authorities also opened a medical institute in Damascus. During World War I, it was transferred to Beirut, and the language of instruction became Turkish. Back in Damascus after the war, the teaching resumed in Arabic, and the scope increased to include law as well as medicine. The evolution of this institution culminated with the establishment of the University of Damascus in 1958.

A few years before World War II, the University of Tehran and the Abadan Institute of Technology signaled expansion of higher education in Iran. While the university covered a broad disciplinary spectrum, the institute of technology, organized by the dominant Anglo-Iranian oil company, concentrated on training petroleum engineers, initially in cooperation with the University of Birmingham.

---

3. Antoine B. Zahlan, *Science and Science Policy in the Arab World*, London, Croom Helm, 1980, 99.

Zionism & Science

Another important means of penetration of modern science into the Middle East in the 20th century was the immigration of European Jews into Palestine. Zionist colonization which began in the late-19th century largely bypassed the existing Jewish communities in Jerusalem and other localities. Mutual mistrust and hostility developed between the old Jewish community and the Zionist pioneers. The latter overtly rejected Judaism and its norms, accepted antisemitic images of the " degeneration " of the traditional Jew and predicated the Zionist enterprise on creating a new " muscular Hebrew ", dedicated to working the soil and building the nation. Their defiant revival of Hebrew — for centuries exclusively " the language of holiness " — as a vernacular further estranged the Zionists from the old Jewish community in Palestine. The estrangement of the two camps endured throughout the 20th century, and the traditional Jew remained largely outside modern science in the Middle East[4]. The situation was quite similar to cultural tensions between scientistic modernizers and traditional Islamic scholars in Egypt and elsewhere in the Middle East. The crucial difference in results was due to the cultural impact of Diaspora Jews who had greatly outnumbered native Palestinian Jews by mid-20th century.

British authorities in Palestine established several institutions of applied character such as a veterinary institute (1925) and a meteorological station (1937). However, Zionism was mostly responsible for developing modern science in Palestine/Israel. The political ideology of Zionism was largely shaped in the period of 1890-1930, *i.e.* during the decades of triumphant scientism. Secular intellectuals were dominant throughout the history of the Zionist movement. Theodore Herzl's *Altneuland*, published at the turn of the century, was a science-fiction rendition of the Jewish future in Palestine. Chaim Weizmann, one of the early leaders of Zionism and later the first President of the State of Israel, was an accomplished chemist. Moreover, his scientific accomplishment in Britain during World War I is often credited as a contributing factor to the Balfour Declaration which articulated Britain's intent to allow " a Jewish homeland " in Palestine. Science and Zionism mutually reinforced each other. It is significant that Albert Einstein was offered the first Presidency of the State of Israel, an offer which he gratefully declined.

The Zionists' attention to science and technology expressed itself in the establishment of the first experimental agricultural station in Central Palestine (1906), the Technion in Haifa (1912) and the Hebrew University in Jerusalem (1924). These were founded at a time when neither the size of the Jewish population of Palestine, nor its industrial needs would justify such a commitment.

4. Yakov M. Rabkin & Ira Robinson (eds), *Interaction of Scientific and Jewish Cultures in Modern Times*, New York, Edwin Mellen Press, 1995.

It was an ideological, visionary gesture funded by Diaspora Zionists which would acquire practical significance only much later. Israeli universities and research institutes developed in subsequent decades became important conduits for modern science into the Middle East.

PRINTING

Besides modern educational institutions, the other route of penetration of modern science into the Middle East was printing. It came to the region in stages. A Hungarian convert to Islam would open a press for use by the Muslims only in early 18th century. The publication of science books in Arabic script began with the introduction of the printing press in Istanbul, long after publication of books in Hebrew and Armenian had blossomed in the same city. Initially, some Islamic circles opposed printing, arguing that the word of God should be only handwritten. The new communication technology was eventually legitimized and spread to other fields, such as science. The Bulak press in Cairo opened in the early 19th century, and by mid-century it had printed nearly a hundred Arabic books on science[5]. A printing press established at the same time in Tehran also introduced elements of modern science in the Middle East.

Western printing was brought to the Middle East by American Protestant missionaries in Beirut in 1834, to be followed by a Jesuit press over a decade later. Educational materials appeared in a variety of fields such as the natural sciences, the arts, philosophy literature, history, and technology. There were dozens of Arabic translations of texts in medicine, chemistry, pathology and physiology printed by the end of the 19th century. Christian missionaries on the staff of the Syrian Protestant College, precursor of the American University of Beirut, played a prominent part in this effort, particularly in introducing scientific literature to the Arabic-speaking audiences.

The opening of printing houses with Arabic script led to an increase in the number of periodicals and scientific journals. Scientific journalism began with the publication of Yacoub al-Tib, a medical periodical, in Cairo in 1865, and Al-Muqtataf in Beirut in 1876. The latter gained prominence as the most important source of reviews of scientific developments in Arabic. It relocated to Cairo in 1885 and was in print until 1952. Al-Muqtataf called for overt embrace of Western civilization, linking the idea of progress with that of evolution. The use of print to promote scientific awareness was supplemented in the 1970s by televised education in science, including open universities that were established in several countries in the region.

---

5. J. Heyworth Dunne, " Printing and Translation under Muhammad Ali of Egypt ", *Journal of the Royal Asiatic Society*, 3 (1940), 325-349.

Governments played the most important role in the development of science in the Middle East. They usually established institutions of higher learning and research, provided most of the budgets, and tried to direct and orient research toward national goals. Modern universities and research facilities in Iran, Iraq, Syria and most other countries of the region were established by governments in the mid-20th century and remained for a long time under different forms of government control. The centralized pattern of science organization in most Middle Eastern countries resulted in strong connections between science and politics. These connections could be beneficial for the development of science but would also make science and scientists vulnerable to political change. Continuity required for science to take root was often absent in a region plagued by political turmoil.

While prior to the Second World War the Soviet Union was the first to formulate and to pursue science policies, the trend became universal, and made its appearance in the Middle East in the 1960s. The tradition of strong government involvement in science prepared the ground for rapid growth of science policy institutions in the region. International bodies such as UNESCO and OECD favored this growth and provided the governments in the region with science policy advisors and standardized science policy indicators and approaches.

RESEARCH CLIMATE

The Middle East produced few scientific achievements of world renown in the 19th and 20th centuries. While there were significant resources invested in expanding scientific and engineering manpower, in building institutions and, particularly, in formulation of grandiloquent science policies, the results were disproportionately meager for most of the region.

A common complaint to be heard from scientists in the Middle East was that their ideas, expertise and energy were underused, that they were "unsolicited" by the economy. Indeed, not unlike the Soviet Union, most countries in the Middle East developed a "hydroponic" science, a science that was poorly rooted in local economy or culture. The fact that the very words "science" and "technology" are absent from the indexes to most bibliographies and economic and political histories of the Middle East, seems to confirm the unrootedness, the irrelevance of science for the region. Except for scientists themselves, science became a going concern for state bureaucrats, sometimes for generals. It remained largely alien for the businessman, for the cleric, and above all for the man in the street and in the market.

This irrelevance contrasts with strenuous efforts made throughout the 20th century to make locally produced science contribute to the welfare of the region. Much research in the Middle East concentrated on practical problems, such as agriculture, water management and medicine. This experience was used in the assimilation of imported technologies, while locally produced research, however valuable, mostly failed to be transformed into technological innovation. Cultural as well as economic factors conspired against innovation based on local ideas just as, in spite of excellent basic research, Soviet economy was dependent on Western technologies throughout the regime's existence. Similarly, industrial companies in the Middle East often preferred to rely on Western laboratories to solve their technical problems, which further accentuated the feeling of irrelevance shared by many scientists in the Middle East. Their research would usually become relevant when technology transfer from abroad was made prohibitively expensive or otherwise inaccessible.

Egypt began with the strongest base for scientific research in the wake of World War II. The National Research Council (NRC) was established in 1939, the earliest such occurrence in the region. The American University in Cairo retained its independence and conducted research, mainly in physics. By mid-20th century most of Egypt's research was done in the government sector which consisted of several modern universities and government research stations, including an entire research campus occupied by the NRC, perhaps, the single largest establishment of this sort in the region. The government modernized older universities (Al-Azhar founded in 930, the Higher Institute of Engineering founded in 1839 and the University of Cairo founded in 1908) and established several new ones.

The bulk of research in Turkey, as in the rest of the region, was concentrated at a few universities and government research centers. The establishment of Tübitak in 1963 created the central body in charge of scientific research in Turkey.

The Pahlavi dynasty strongly encouraged scientific research in Iran. Science was meant to modernize Iran, it also provided prestige and became a status symbol for the dynasty. Military cooperation with the United States and Israel also stimulated Iran's scientific and technological activities. The Pahlavis found legitimacy in a commitment to modernization, combined with references to Iran's pre-Islamic past. While they strongly encouraged modern science and education, the Pahlavis, unlike Atatürk and his heirs, did not destroy or even control the Islamic educational infrastructure which would foment revolution against the Shah in the late 1970s in which large numbers of engineers also took an active part.

The fall of the Shah in the course of the Islamic Revolution was initially seen as a blow to the country's modernization. However, subsequent Islamic governments fully recognized the importance of science and largely continued previous governments' policies with respect to local R&D. The new system of *Jihad-e-Daneshgahi* (University of Jihad) oriented research toward practical goals, renamed universities, and encouraged publication, all with the express purpose of placing science in the framework of Islam. Decrease of international contacts, in the aftermath of the revolution and in the course of a protracted war with Iraq, inhibited research in the first post-revolutionary decade. Iran, as well as Iraq, was also affected by Western measures aiming at non-proliferation of weapons of mass destruction.

Iraq began scientific research activity in the 1960s. University of Baghdad was established in 1957 on the basis of several colleges organized since the early 20th century. In turn, it engendered several universities in the provinces in the 1960-1980s. The Iraqi Atomic Energy Commission, established in 1956, developed the country's most active research establishment under its auspices, offering technical services in several industrial and agricultural fields ; it conducted thousands of tests in 1972 in the fight against a mass mercury poisoning in the country. Iraq became particularly active in weapons research, using its impressive oil revenues for this purpose.

Oil revenues were also used in other countries to develop scientific research. Saudi Arabia and Kuwait made a strong commitment to research which was initially conducted by expatriate staff, including many of Egyptian, Lebanese and Palestinian origin. Research topics were largely formulated in response to local needs rather than to transnational trends of fundamental inquiry.

Syria's research was concentrated in several universities, the most established of which was the University of Damascus. The newer University of Aleppo developed better research facilities, including the Institute for the History of Arabic Science investigating classical science. Syria hosted several regional institutes of research on arid zones, a subject of great concern and relevance. However, for political reasons Israel, with its unique expertise in the field, was not invited to join.

Israel, and its science, developed largely in isolation from the region and, for most of its history, in military conflict with most Arab states. Besides the military sector, Israel developed a strong tradition of basic research, financed by government grants as well as by important private funds raised abroad for specific programs and institutions. Israel's R&D expenditure as percentage of the GNP exceeded those of smaller European countries and Canada in the late 20th century[6].

The State of Israel experienced a surplus of scientific and technical manpower during most of its history. This reflected the prominence of Diaspora Jews in world science, and the return of Israeli-born Jews to Diaspora patterns

6. Asher Irvin *et al.* (eds), *Strategies for the National Support in Basic Research : an International Comparison*, Jerusalem, The Israel Academy of Sciences and Humanities, 1995.

of career choices long objected to by the Zionist pioneers. Israel also received tens of thousands of research-front scientists as immigrants, particularly, from the USSR and the successor states. A scientific vocabulary in Hebrew was consistently developed which facilitated the native Israelis' access to science. Attempts at Arabisation of science, started in the early 20th century in several countries, were less conclusive. The introduction of Persian as the scientific language of Iran came to be debated only in the end of the 20th century.

SCIENTIFIC COOPERATION

Peripheral countries are not known to develop strong scientific links among themselves, and usually rely on established networks covering mostly the developed countries. In this sense, the Middle East follows the trend, perhaps, with the exception of Israel. In order to compensate for its isolation in the region, Israel established strong research ties with major Western countries. Cooperation with, and financing from, the Unites States and Germany played a crucial role in fomenting basic research in Israel. Whenever politically feasible, Israel developed scientific and technological cooperation with less developed countries, including the pre-Khomeini Iran. It also cooperated with other regionally isolated countries such as South Africa in the 1970s and 1980s.

Israel's cooperation with Arab countries, initially with Egypt, began in the 1980s. It was largely financed and politically encouraged by the United States, in line with its policy of reconciliation stemming from the Camp David Accords. In 1982 the Israeli Academic Center was opened in Cairo. Most cooperative projects between Israel and Arab states centered on medical and agricultural research. Benefits of this cooperation were of both political and scientific nature. Jordan developed cooperative research initiatives with Israel in the 1990s.

While Israel had political reasons to pursue scientific cooperation with Arab states in order to gain recognition and normalize relations, Turkey had no such concern, and it concentrated most of its international contacts outside of the Middle East. The political, military and technological rapprochement with Israel in the 1990s created a favorable basis for scientific cooperation between the two countries.

Republican Turkey developed science and technology cooperation with Russia and Ukraine even before the formal establishment of the USSR in 1922. In the last decade of the 20th century Turkey gained broad access to post-Soviet science and technology, especially in the Caucasus and in Central Asian republics whose cultural affinity with Turkey greatly facilitated this cooperation. Scientific relations with Central Asia were also pursued in concert with Iran and Pakistan in the framework of a regional organization for economic cooperation. However, it was Israel that derived most significant benefits from Soviet-trained researchers in the last three decades of the 20th century.

In the context of Arab nationalism fomented by President Nasser, the government expanded scientific and engineering training while, at the same time, it provoked an exodus of non-Arab talent, including researchers, from the country. As Egyptian society grew less cosmopolitan after 1956, the government pushed for regional, or rather Arab, science initiatives. Several institutions, including a regional Arab center for radioisotopes and a center for regional studies, resulted from this policy. The creation of the League of Arab States in 1945 led to the establishment of the Arab League Educational, Scientific and Cultural Organization (ALESCO) in 1970. ALESCO promoted scientific and technological cooperation among Arab governments, and in 1977, it started a science education program. The foundation of the Arab Federation of Scientific Research Councils (FASRC) was meant to enhance research cooperation among the Arabs. However, according to Antoine Zahlan, a veteran observer of science in the Middle East, Arab unity, in spite of its rhetoric, remained irrelevant to the progress of science in the Arab states[7].

Turkey was never part of an Arab framework. Atatürk, an ardent adept of scientism, considered science " the best guide to the world ", and he single-handedly encouraged science and higher education in Turkey. Atatürk presided over a major reform of science that was undertaken with the help of German, mostly Jewish, scientists fleeing the Third Reich. These émigrés made a major contribution to the development of science and higher education in the Middle East, in Turkey as well as at Jewish institutions of the British Mandate Palestine and, later, of Israel.

Bilateral cooperation with the United States, with the Soviet Union and with the European Union and its members involved most countries of the Middle East, with Egypt, Israel and Turkey being the most active partners. Turkey was also an active partner of scientific cooperation in the NATO framework, of which it alone in the region was a full member.

The Organisation of Islamic Conference encouraged scientific cooperation that involved all the countries of the Middle East, except Israel. Alongside with sponsorship of research in science and engineering, it also established an international center in Istanbul which, since the 1980s, heightened the public awareness of science and the history of science in the region. As a result of this initiative, there developed a better understanding of the origins of modern science in the Middle East, particularly of its cultural aspects. Research on the

---

7. Antoine B. Zahlan (ed.), *The Arab Brain Drain*, London, Ithaca Press, 1981, 17.

history of Ottoman and Turkish science was also conducted at the Technical
University in Istanbul and at the University of Ankara. While research in Turkey and Syria primarily dealt with science in the region, the history of science
developed in Jerusalem and Tel Aviv reflected world trends, and only by the
end of the century turned to the history of Jews in science, long considered
" too parochial " a topic by Israeli academics yearning international recognition.

RESEARCH PERFORMANCE

Publications in international journals[8] can be used as a criterion of research
performance in the region, or at least as a criterion of integration of Middle
Eastern science in international networks of scientific communication. In terms
of publications, the region rated rather low on a per capita of researcher basis.
Internationally visible publications concentrated mostly in the sciences, while
research in social sciences and humanities rarely spread beyond national borders. By the end of the 20th century, the share of the social sciences and the
humanities in total publication aggregates was between 12 and 15% for Israel,
Jordan, Kuwait and Lebanon ; however, in Turkey this share was 5%, in Egypt
3%, and in Iran 2%.

Within the natural sciences drastic disparities could also be found between
countries of the region. In the 1970s, Israel's publication output per capita of
population was the highest in the world : a hundred times higher than that of
its Arab neighbours, and almost twice as high as in the United States. Egypt's
scientific output was about a quarter of that of Israel in the 1970s and 1980s,
and declining in relative terms to less than one fifth in 1998. Egypt published
twice as many scientific papers as Turkey in 1985. However, by the end of the
century the positions were reversed : Egypt's production was just 44% of Turkey's. The cumulative Arab Middle Eastern scientific output as a share of
Israel's output also declined in the same period to about one third by the end
of the century. At the same time, there was notable progress made in the number of publications in the oil producing countries with the exception of Iraq
whose contribution to world science declined as a result of the Gulf War and
the ensuing problems (from 257 publications in 1985 to just 34 in 1998).

Important research, particularly in biomedical sciences, was conducted at
the American University in Beirut in the 1950s-1970s. Lebanon also showed a
different, *i.e.* non-governmental pattern of the development of science. Civil
war and the Israeli invasion made a dent in Lebanon's scientific growth but the

---

8. The data come from the Institute for Scientific Publications (ISI), USA.

production of articles reached its pre-war levels and surpassed it by the early 1990s.

Responding to strong government incentives, Turkey's research steadily increased its world visibility, positioning it a strong second to Israel (3901 vs 9544 articles in 1998). The scientific prominence of Israel and Turkey may be historically explained by the crucial role played by émigré German Jewish scientists in the development of science in these countries in the 1930s, and the continuing reliance of Israel's science on human and financial resources of the Diaspora.

However, alongside the disparities, there were also some similarities among countries of the region : a disproportionate share of internationally visible publications produced by Arab, Iranian, Israeli and Turkish scientists alike resulted from research trips abroad and cooperative projects with Western colleagues. Moreover, foreign-trained scientists tended to publish more frequently and in more prestigious journals. In terms of impact on world science measured in citations, Israeli scientists were ahead of all its neighbors, and even ahead of the rest of the world in such fields as Computer Sciences. But the region's overall research visibility resembled that of the Soviet-bloc countries with their impressive personnel figures and relatively little effect on world science. This could be interpreted in a variety of ways : inadequate working conditions for the scientist, inappropriate training, cultural and political isolation, concentration of research on non-publishable, mainly military projects.

THE MILITARY

Indeed, an important stimulus that selectively propelled scientific development in some countries of the Middle East (Iran, Iraq, Israel, Turkey) was of military nature. Turkey was integrated in the NATO procurement policies and benefited from technology transfer of Western provenance. Availability of scientific and engineering manpower, some of it trained in the West, facilitated this transfer and, in turn, stimulated scientific activity and advanced training in the country. Israel also benefited from NATO procurement needs in the 1980s and 1990s but would find itself at odds with the United States and, earlier, with France with respect to its own military purchases in these countries. Israel found itself in the company of Iran and Iraq which were denied certain kinds of weapons, and these three countries opted for designing and producing them in the 1960s-1990s. They had to rely mostly on local human and material resources to substitute for imports, and only rarely resorted to the services of foreign scientists and engineers.

Weapons of mass destruction, all of them science-based, figured prominently in the list of such denied imports. As early as in 1960 President Nasser of Egypt warned that " if it is confirmed that Israel is manufacturing a nuclear weapon, then this would constitute the beginning of war, for it would be imperative that we attack this base of aggression ". While Israel was apparently the first to acquire nuclear weapons in the region, Israel also turned Nasser's logic against Iraq and bombed its Osirak reactor, suspected to develop nuclear weapons, in 1981. In the following decade, major Western powers used both on-site inspection and air strikes to impede research and production of weapons of mass destruction by Iraq. Isolated bombardments of other suspected science-based weapons facilities took place in Libya, Sudan and Afghanistan in the 1980s and 1990s. This could also be seen as *sui generis* recognition of science and technology development in these countries.

Governments in the Middle East not only built infrastructures for science, they remained the single most important promoter of scientific research throughout the region. For example, fiscal incentives encouraged Israeli scientists to take sabbaticals abroad while government subsidies enticed Turkish scientists to publish in internationally recognized journals. Major government efforts were invested in making science relevant to industrial concerns. These policies were predicated on the common belief that scientific research was the main motive force of industrial innovation.

However, industrial applications of science remained weak in all the countries of the Middle East. Adoption of economic liberalization measures and the cancellation of a major military aviation project (leading to massive, albeit well compensated lay-offs of R&D personnel), created conditions for the science-based entrepreneurial activities in Israel to blossom in the late 1980s-1990s. It is significant that this successful integration of science and business was a side effect of an otherwise decried decision to cancel the development of a local jet fighter.

Turkey's economic presence in the Middle East in the fields of construction and production of electric appliances relied little on Turkish R&D. By the end of the 20th century, less than one percent of Turkey's manufacturing companies conducted any R&D work. The scarcity of venture capital was another inhibiting factor mitigating against science-industry links throughout the Middle East. For many years it was an important bottleneck even for Israeli companies which developed strong links with the world's financial markets only in the 1990s.

Most countries of the Middle East experienced substantial brain drain that included scientists settling in Western countries, many remaining there after completing their training abroad. The problem attracted attention in the 1950s-1970s as a sign of inadequacy of local R&D opportunities even though there was no shortage of scientists reported in the " donor " countries. Ritual complaints about brain drain from the region dominated public discourse at the expense of empirical studies of actual needs and opportunities in the more developed countries of the region. International prominence of Arab, Israeli or Turkish émigré scientists was often lamented as a loss to the countries of the Middle East. The oil boom of the 1970s changed the situation and attracted some of this talent back to the region, particularly to universities and research centers in Kuwait, Saudi Arabia and other oil-rich countries. The pool of professionally productive expatriate scientists and engineers was also used as a source of culturally appropriate technical expertise. For example, Turkey involved non-resident Turkish scientists in directing the country's solar and nuclear energy projects in the late 1970s.

SCIENCE AND TRADITION

Modern science in Europe was often presented as an alternative, sometimes even a substitute for religion. The history of conflict between religion and science is well publicized, even though a revision is currently under way which seems to suggest the existence of ideologically motivated distortions. Whatever its relevance for Europe and the Americas, the history of conflict is quite limited in the Middle East, because of important cultural and chronological differences.

Since Islam and Judaism have no established institutional centre (such as the Vatican), their interpretation of the respective sacred texts is contingent on tradition. It may be more appropriate to consider tradition, *i.e.* the way science is presented in educational and other frameworks, rather than religion, as a factor in the development of science in the Middle East. The import of the opposites " religion and secularism " into Arabic and Hebrew occurred precisely at a time when science was also being transferred from the West. While the Hebrew word for secularism (*hilun*) has no scientific connotation, the Arabic term ('*ilmaniya*) is derived from '*ilm*, the Arabic for science, and creates the impression that the main goal of secularism is the development of science. Some scholars objected to this usage in the late 1970s since it separated Islam from science. They came up with another term for secularism (*al-ladiniya*), literally " irreligiosity ". This debate points out at the limitations of using these Western concepts in analyses of the place of science in Islam and Judaism[9].

In both Islam and Judaism borders between science and tradition were often quite blurred. The tradition concerned itself with many an aspect of life and its

9. See : B. Tibi, *The Crisis in Modern Islam*, Salt Lake City, University of Utah Press, 1988, and Yakov M. Rabkin, " Conceptual Issues in the History of Science in non-western Societies ", in Ekmeleddin Ihsanoglu (ed.), *Transfer of Modern Science and Technology to the Muslim World*, *loc. cit.*, 59-66.

cognition that, for the Christians in the 19th and 20th centuries, lay outside the domain of religion as they had come to define it. Both science and tradition develop through transmission (albeit in different manners) and both are essentially cumulative. But they do differ with respect to the source of knowledge and, most importantly, to the legitimacy of its modification.

Modern science attracted Middle Eastern elites mainly because of its extrinsic rather than intrinsic value. Technology, not disinterested advancement of human knowledge, became the real objective of scientific development, and in this sense the growth of science in the Middle East was not accompanied by the same ethos that lay at the root of modern science in Europe. Moreover, reception of science and reception of technology provoked different responses in the Middle East. Opposition to certain models of cosmology could be observed alongside with wide-spread uses of modern communication technologies by devout Muslims and Jews.

As Middle Eastern countries embarked on development of science, the attitudes of Christianity, Islam, and Judaism towards science were vigorously debated throughout the last two centuries. One may identify " four main positions " articulated with respect to relations between science and tradition, an issue that does not seem to fade with time. The first one accepted the conventional view that science is incompatible with religion. It sometimes included the belief, borrowed from the West, that the advancement of science was predicated on the liberation of human reason from the shackles of religion. This was the militant opinion voiced by such national leaders as Kemal Atatürk and David Ben-Gurion, both of whom practiced enlightenment by force, and created a durable image of science as an antithesis to religion. Christian Arab thinkers, such as Shibli Shumayil, also argued for a new science-based culture in the Middle East and were the first to spread Materialism and Darwinism in the region in the late 19th century. Muslim critics have summarily traduced such scientistic approaches and dubbed them " westoxification ".

A second, less categorical position acknowledges the transnational character of scientific knowledge but postulates that tradition should impact on research priorities and orientation. In opposition to the first position, it was stated that " a nation that imports 'modernity' pays a high price : it is cultural suicide ". Modernity must be elevated, sanctified, and, to be properly assimilated, has to be integrated into the traditional normative framework. Science also had to be submitted to moral controls derived from tradition. According to this approach, all knowledge ultimately had to be focused on monotheistic principles and ideas.

The major Islamic scholar Al-Afghani wrote in the second half of the 19th century that Islam was compatible with science which he deemed to be " a noble thing that has no connection with any nation ". His conclusion influenced the minds of many a Muslim thinker in the 20th century who argued that cultivation of science was essentially a Muslim duty[10]. Ayatollah Mutahari (Iran, 1970s), declared scientific education to be a *jihad* under the command of Islamic values. Sufis found little difficulty in integrating Einstein's theories within their conceptual framework, and they hailed and praised him in many different ways, such as Islamic poetry composed to honor Einstein and his view of the universe[11]. Utopias written in the Middle East also suggest that science and technology had become part of Islamic images of the future (e.g., association between the space shuttle and the Prophet's ascent)[12]. There were also claims from scientists, both Jewish and Muslim, that their cultural background enriched their scientific work.

It is with the goal of achieving an overarching holiness that science was enthusiastically included in Judaic curricula by Rabbi Samson Raphael Hirsch (Germany, mid-19th century), whose influence could be felt for more than a century in Israel (e.g., at the Jerusalem College of Technology) and elsewhere among Orthodox Jewish scientists[13]. Adepts of this position would not argue that revealed knowledge was the sole basis of understanding man and his origins. Their reactions were usually more nuanced. They would argue that understanding the world is an imperative, both for pragmatic (be they technological or medical) reasons and for the purpose of glorifying one's religion as a religion of knowledge and understanding. Quite in concert with Ayatollah Mutahari, Rabbi Yosef Kafah, a prominent contemporary scholar and adjudicator, considered acquisition of scientific knowledge a divine commandment, and strenuously objected against cultivation of scientific ignorance in some Jewish schools.

When "undesirable knowledge" had to be dealt with, a variety of ways were designed to question its validity. Within this position, emphasis is put on a relativist approach which considers science a product of its time, and therefore apparent contradictions are taken in stride. Rabbi Kafah stated that scientific knowledge contained in the Talmud and other normative Judaic texts should be treated as historically contingent and therefore by no means binding. Similarly, the late-20th century Iranian intellectual Abdolkarim Sorush demystified Darwinism with the help of Karl Popper's philosophy of science, and

10. Farhang Rajace, " Islam and Modernity : the Reconstruction of an Alternative Shi'ite Islamic Worldview in Iran ", *Fundamentalisms and Society*, 103-126.

11. Thiery Zarcone, " Une récupération d'Einstein chez quelques auteurs mystiques du monde turco-iranien ", *Turcica*, 21-23 (1991), 131-154.

12. Christian Czyska, " On Utopian Writing in Nasserist Prison and Laicist Turkey ", *Der Welt des Islams,* 35(1), 1995, 95-125.

13. Leo Levi, *Torah and Science : their Interplay in the World Scheme*, Jerusalem, Feldheim, 1983.

presented it as a mere tautology. Adepts of this position actively pursued hybridization of local traditions with scientific education, largely undermining the essentially positivist state systems of education.

The third position goes beyond emphasis on temporal relativism, and does not regard science as transnational : science is viewed as a Western cultural construct in both style and content. Yet, it also argues for its harmonization with Muslim knowledge. Mahdi Bazargan, a mid-20th century Iranian states-man and educator, denied that " the sun of science is shining from Western Europe ". He also showed that modern scientific concepts could be found in Koran. A cursory search for the combination of words " science " and " Islam " reveals a plethora of sites trying to show how Koran had foreseen most impor-tant scientific discoveries. In the same vein, devout Jews spare no efforts to find Biblical and Midrashic confirmations of major scientific discoveries and modern technological artifacts. Such conciliatory movements could be found in the Middle East throughout the 19th and 20th centuries. Science was thus val-orized with the help of canonical texts (even though some may suspect that the traditional texts were thus also valorized by science). In educational institu-tions reflecting this approach, sciences have to be taught by devout adepts of Islam and Judaism.

The fourth, most radical position, suggests that Islamic science should be based on concepts of man, nature, space and time derived from the tradition rather than from Western science. This would be a science quite different from the rest of world science. The latter position offered fruitful ground for debate but was not realized in practice. By the end of the 20th century, an active inter-national network of scientists and philosophers, including those in the Middle East, made attempts to define science in Islamic terms and to make it part of the newly assertive Islamic culture[14]. However, there were no crude attempts to subjugate science to ideological diktat resembling the story of " German physics " or " Proletarian Biology " in mid-20th century Europe. Rather, it was postulated that all modern science was derived from Islamic science, and that the Muslims must simply " repossess " science as part of a larger movement of the " return to the self ". While there were similar attempts made in the world of Hinduism (notably by the physicist J.C. Bose) there seems to be no known Judaic variant of this approach. One of the reasons may be the sad memory of serious attempts made by anti-Semitic scientific luminaries in Germany, including Nobel laureates, to denounce " Jewish science " as an antithesis to the allegedly healthier and more intuitive " Aryan science " to be pursued by true National-Socialist Germans.

14. Ziuaddin Sardar, *Explorations in Islamic Science*, London, Mansell, 1989.

Beyond these four approaches that actually engage science there exists a trend which opts to ignore it altogether. It postulates that science is not only Western and, unlike Torah or Koran, time-bound but it is also superfluous since all scientific knowledge can be found in the canonical texts. Stories abound of rabbis and imams proffering scientific and medical information without ever opening a book on science or medicine. Much use is made of the above-mentioned attempts to find confirmation of scientific discoveries in traditional commentaries to Torah and Koran. Consequently, educational institutions based on these principles do not teach science at all or keep it, mostly mathematics, to a bare minimum needed to run a small shop. Quite a few schools ostensibly teach science to satisfy government requirements but, in essence, consider it useless. At such schools scientific results that appear to contradict traditional views would be simply ignored. The school administrators would censor textbooks, purge " seditious " scientific ideas, often by tearing out entire chapters or just blackening offensive passages.

For opportunistic purposes, it was not uncommon in the Middle East in late 20th century to make references to Koran in scientific publications in order to secure lucrative teaching positions at universities in Saudi Arabia and other Islamic countries. The situation was different in Israel and Turkey where those committed to the secular character of the state grew defensive with respect to ostentatious religiosity. Secularists' protests about the emergence of a new Torah-oriented school network for Sephardim in Israel and the prohibition of head covering for female students at Turkish universities were instances of such defensiveness in the 1980s and 1990s.

Attempts to oppose Islam to modern technology usually failed. When Saddam Hussein proclaimed a slogan " Islamic belief against technology " in the context of the 1991 Gulf War, it was seen as a sign of impotence rather than devotion. Governments in the Middle East uniformly developed a pragmatic attitude to science, whatever their degree of identification with Islam. However, by the end of the 20th century, various degrees of estrangement from science could be observed throughout the Middle East. Tradition and science might no longer be antagonists but for most inhabitants of the Middle East they remained culturally unrelated and alien.

CONCLUSIONS

Modern history in the region is characterized by almost exclusive reliance on the state, often on the military. Private institutions are largely insignificant, and to the extent they serve as loci of research they are often of foreign provenance. While there were several attempts made to foster scientific cooperation within the region, its importance was negligible in comparison with the vital contacts scientists in the Middle East maintained with their colleagues in Western countries. It is these foreign contacts that accounted for a lion's share of the region's scientific production. All of these factors tend to isolate scientists from local population largely imbued with traditional beliefs and practices. There were several attempts made to integrate scientific pursuits with the traditional values of Judaism and Islam, and these attempts are likely to continue with the regain of religiosity observed in much of the region. However, at this writing, most of the science done in the region remains marginal with respect to both local culture and to world science.

BEYOND CULTURALISM ? AN OVERVIEW OF THE HISTORIOGRAPHY ON OTTOMAN SCIENCE IN TURKEY

Cemil AYDIN

During the twentieth century modernization in the Muslim world, history of science gained far more significance in historical consciousness beyond the scholarly interest in understanding the past of science and scientists. While Muslim societies were transferring Western science, the reasons that led to the gap in the scientific development of Islamic world and Europe became both a scholarly question and an ideological controversy. Especially in modern Turkish Republic history of science became an essential part of the official nationalist ideology and cultural identity. In the context of the late Ottoman and early Republican ideological debates, different ways to explain why Ottoman Muslims lagged behind Europe were closely connected to different reform agendas.

The extraordinary importance of history for the legitimization of competing visions of reform and modernity became both a blessing and a disadvantage for the historiography on Ottoman-Islamic science in Turkey. This paper presents an overview of this historiography to discuss the positive and negative impacts of the particular ideological entanglements of the history writing. It argues that a grand question of what to blame for the " decline " of Ottoman power in comparison to Europe as well as a linear vision of modernization and progress led to a very essentialized and polarized debate about the relationship between Islamic religious tradition and modern science. Even the two leading historians of science in Turkey, namely Adnan Adıvar and Aydın Sayılı, restricted their research agenda and questions to the formula of Islamic faith against modern science. As a result, historiography on the Ottoman science lagged behind for a while compared to methodological innovations in the discipline. However, recent critiques of the legacy of the culturalist debate on the relationship between " Religion and Science " led to a sociological agenda of research on Ottoman science, particularly in the works of Ekmeleddin Ihsanoğlu. This new research agenda on Ottoman science could now turn the negative legacy of the over-discussed theme of religion and science into an advantage to offer not only new methodologies to explore the question of cultural diversity and modern science, but also ways to overcome the ideological polarization around the issues of modernity and Islamic tradition.

Emergence of Two Competing Culturalist Visions of Ottoman Science

Ottoman reformers have long reflected on the role of science and technology as important factors that led to the rise of European economic and military power in comparison to the Ottomans. Initially, Ottoman reformers did not consider Islamic culture and religion as the major factor neither in the relative decline of their power nor in the success of the reform projects. However, by the late 19th century, Islam as a religion began to be discussed as the basic unit of analysis to explain the rise or decline of science among Muslims.

Two main controversies regarding the relation between science and religion during the 1880s and the 1890s might indicate the genealogy of the two conflicting interpretations of the rise and decline of Islamic science, which takes Islamic culture as the basic explanatory unit. First one is the discussion around the translation of John William Draper's *History of the Conflict Between Religion and Science* into Ottoman-Turkish[1]. Draper's main contention was that an ongoing war between science and religion had been waged throughout history. Although the translator Ahmed Midhat wrote a critical preface to the translation arguing that Drapers's argument is not true for Islamic religion if it may be true for Christianity, as Adnan Adıvar recalls, the main text had more influence on the readers than the apologetic notes of the translator. The other significant event was the Ottoman responses to the Ernest Renan's speech entitled " Islam and Science " (*L'Islamisme et la Science*) given in Sorbonne in 1883[2]. Renan's famous argument about the inevitable and essential conflict between Islamic religion and modern science contained not only explicitly racist arguments about the incapability of Semitic/Arab and Turkic races in science, but also a conviction that religion in general and Islam in particular is hostile to science and progress. Renan argued that the past scientific achievement in the Muslim world was not " due to Islam ", as on the contrary, it occurred " in

---

1. John William Draper (1811-1882), *History of the Conflict between Religion and Science,* England, Gregg International Publishers, 1970. (Reprint. Originally published in London by Henry S. King in 1875). For the Ottoman translation of Draper's works, see *Niza-yi Ilim ve Din*, translated and introduced by Ahmet Midhat, Tercüman-ı Hakikat Matbaası, Dersaadet (Istanbul), 1313 (1896 or 1897). For a contemporary critique of Draper's views, see Donald Fleming, *John William Draper and the Religion of Science*, Philadelphia, University of Pennsylvania Press, 1950.

2. For the original text of the speech, See, Ernest Renan (1823-1892), *L'islamisme et la science : conference faite a la Sorbonne le 29 mars 1883*, Paris, Calmann Levy, 1883. For a good Turkish translation : Ernest Renan, " Islamlık ve Bilim ", *Nutuklar ve Konferanslar*, transl. Ziya Aslan, Sakarya Basımevi, Ankara, 1946. For the major views of Renan on Science in English, see the translation of his *Avenir de la science, The Future of Science*, Boston, Roberts Brothers, 1891.

spite of Islam " and it was done not by Arabs, but non-Arab Aryan Iranian race and Christians[3]. Therefore, according to Renan, historical achievements of Muslims in sciences do not contradict with the " science vs. religion " principle, since this success has nothing to do with the influence of Islamic faith. Muslim intellectuals of the time wrote refutations to Renan's arguments. However, responses to Renan's speech, Draper's work and other similar works demonstrate the saliency of the issue of Science and Religion among the Ottoman intellectuals, and gradually influenced the way Ottoman intellectual interpreted the Ottoman backwardness in science, especially in the context of the transition from Ottoman State to the Republican Turkey.

It was a combination of several factors that shaped the views that considered Islamic culture as the major reason for the " backwardness of science " among Ottoman Muslims. Influence of materialistic and positivistic intellectual currents, the failure of earlier piecemeal reform efforts by Ottoman leaders, conflicts between Western style school graduates and *madrasa* graduates impacted the social background of the Islam-negative culturalism. Against this argument, Islam-positive culturalists claimed that Islam was the main impetus for the rise of Islamic science and only the deviations from " true Islam " can be responsible for its decline. This argument was initially a nationalistic theme to counter Orientalist arguments that relegate Muslims to an inferior position not only in contemporary politics, but also in world history, but gradually turned into a main pillar of Islamic modernism. There was a third group : those trying to employ Islam just as a modernizing tool, as part of cultural nationalism and confidence[4]. Needless to say, to draw clear cut lines among these three groups is never easy, and boundaries are always blurred. Yet, in the complexity of Turkish intellectual life from the time of Abdulhamid II to the Republican period, the question of the relationship, either negative or positive, between Islam and modern science became a controversial and popular topic.

---

3. Expectedly, there were immediate reactions from Muslim intellectuals, three of whom are well known : Jamaladdin Afghani, Namık Kemal (Istanbul) and Ataullah Bayezidof (Mufti of Russia). For Namık Kemal's response, see, Namik Kemal, *Renan Müdafaanamesi : Islamiyet ve Maarif*, Ankara, Milli Kültür Yayınları, 1962. For Afghani's response, see, " Answer of Jamal ad-Din to Renan ", in Nikkie Keddie (ed.), *An Islamic Response to Imperialism : Political and Religious Writings of Sayyid Jamal al-Din al-Afghani*, Berkeley, University of California Press, 1968. For the response of Ataullah Bayezidof, see, Ataullah Bayezidof, *Islam ve Medeniyet*, Ankara, TDV Yayınları, 1993. Sayyid Amir Ali, Rashid Rida, Celal Nuri and Muhammad Hamidullah were other Muslim intellectuals who wrote refutations to Renan's arguments. For a bibliographic survey on the refutations of Renan, see Dücane Cündioğlu, " Ernest Renan ve 'Reddiyeler' Bağlamında Islam-Bilim Tartışmalarina Bibliyografik Bir Katkı ", *Divan*, Vol. 2, Istanbul, 1996, 1-94.

4. See Şükrü Hanioğlu, " Osmanlı Aydını ve Bilim ", *Toplum ve Bilim*, n° 28, Güz, 1984. Also Şükrü Hanioğlu, *The Young Turks in Opposition*, New York, Oxford University Press, 1995.

" Islam-negative " Culturalist Historiography on Ottoman Science
Adnan Adıvar and History of the (Absence) of Ottoman Science

Adnan Adıvar was the first intellectual historian who wrote a systematic account of history of science among Ottoman Turks. As a military-medical doctor who had witnessed the final disintegration of Ottoman State, Adıvar's work represents, to a large extent, the general climate of opinion among his generation regarding the relationship between the long-term cause of Ottoman decline and the history of Ottoman science. We can read the angry mood, frustration and inclination to blame something in history for the failure of the Ottoman Turks to catch up with the modern societies of the West in Adıvar's work *La science chez Turcs Ottomans* (Science among the Ottoman Turks)[5]. His main argument was that Ottoman Muslim society became inhospitable to the flourishing of science and technology due to the Islamic dogmatism of their culture. In this inability to advance in science and technology, Ottoman intellectuals must be blamed more than other non-Western intellectuals, because Ottomans were in a favorable position to be in contact with Western societies[6]. But, Adıvar argues, Ottomans were not even aware of the great scientific achievement of Europe during the scientific revolution, due to a civilizational barrier that was created by Islamic dogmatism of the Ottoman religious institutions and the ulema. In his historical account, Adıvar talks about several free souls of Ottoman intellectual life, which includes Katip Çelebi or Ottoman Sultan Mehmed II, who had realized the significance of science and interaction with Europe. Yet, in Adıvar's historical narrative, all of those free souls were suffocated by the fanaticism and indifference of their society. Adıvar offered comparisons of the scientific activities in the Ottoman world and Europe in each century, a comparison in which the Ottomans were always inferior[7].

Adıvar is equally harsh in his critics of the reform efforts of the Tanzimat period. After citing the late 19th century Ottoman intellectual Ahmet Mithat's affirmation that there is nothing contrary to science in Islam, Adıvar surprisingly concludes that " Incidents such as this strengthen one in the conclusion that most of the Turkish intellectuals of the Tanzimat period retained their Oriental mentality and culture, with all the associated and antiquated beliefs,

5. Adıvar completed his book first in French in France, when he was an *émigré* from Republican Turkey, and a revised and enlarged translation of his work appeared in Turkish only in 1943. See, Abdülhak Adnan, *La science chez Turcs Ottomans*, Paris, 1939 ; and Adnan Adıvar, *Osmanlı Türklerinde Ilim*, 4th ed., Istanbul, Remzi Kitabevi, 1982.

6. Abdülhak Adnan Adıvar, " Interaction of Islamic and Western Thought in Turkey ", in T. Cuyler Young (ed.), *Near Eastern Culture and Society : A Symposium on the Meeting of East and West*, Princeton, Princeton University Press, 1951.

7. Adnan Adıvar, *Osmanlı Türklerinde Ilim*, *op. cit.*, for some examples, see, p. 54-57, 121-125 and 177-180. Bernard Lewis, underlined the same theoretical conclusions with more historical detail in a book appropriately dedicated to Adnan Adıvar. See Bernard Lewis, *Muslim Discovery of Europe*, London, Weidenfeld and Nicolson, 1982.

while adopting the technical side of modern life "[8]. What is more surprising is Adıvar's judgment on the Republican period, during which he saw a continuation of the same backward " oriental mentality " in a new positivist garb. Since in the first 25 years of its life, Republic did no better than its Tanzimat ancestors, Adıvar again attributes this failure to " Oriental mentality and culture " and the fanaticism of the new positivist religion[9].

The strong dose of finding blame in cultural traditions and mentality, which sometimes reminds the style of Renan with concepts such as " Oriental mentality ", can be interpreted as the result of Adıvar's frustration with the failure of several reform attempts during both Tanzimat and Republican periods. He seems to find a scapegoat for the decline in concepts such as " dogmatism ", " mentality ", " fanaticism ", which have little historical contextualization throughout the book. His constant comparisons with the European case, his total disregard for the sociological or global causes for the decline of Ottoman power, and his imaginative and dramatic narrative of presenting " good people " who were suffocated by the bad ones render his book a very gloomy and pessimistic one. Adıvar saw the disintegration of the Ottoman Empire during the World War I, during which he served the Empire as a medical doctor. His book gave the final judgment that Ottoman Empire and its civilization deserved to collapse long before the World War I due to Ottoman ignorance of science and technology, especially its unawareness of the great scientific developments in neighboring Europe.

Aydın Sayılı and " Scientific Revolution " Centered Ottoman History

Aydın Sayılı, whose dissertation at Harvard University on the institutions of science and learning in the Muslim world was the first Ph.D. dissertation in the academic field of history of science, wrote extensively on Ottoman science[10]. His research findings proved that Islamic astronomy, even if not the whole scientific dynamism in Islam, continued to develop up to the middle of the 16th century, and there were strong influences on the Copernican theory from 14th

---

8. A.A. Adıvar, " Interaction of Islamic and Western Thought in Turkey ", *op. cit.,* 124.

9. *Idem,* 128 : " The domination of Western thought, or rather of the positivism of the West, was at that time (Republican period) so intense that one can hardly call it thought. It should be termed rather the official dogma of irreligion... Within the last twenty years the vast majority of Turkish youth has been brought up without any official religious teaching, Western positivism being imposed on it just as Islamic dogma had been imposed in the past. At the present moment the New Thought has assumed much the same position as was formerly occupied by the old Islamic dogma ".

10. George Sarton was the supervisor. Part of his Ph.D. thesis, " The Institutions of Science and Learning in the Muslim World ", was converted into a book, see, Aydın Sayılı, *Observatory in Islam,* Ankara, Türk Tarih Kurumu Basımevi, 1960.

and 15th century Muslim astronomy[11]. Moreover, Sayılı's research concluded that Islamic astronomy came very close to Copernicus and provided most of the astronomical data for Copernican Revolution. This led him to formulate a Needhamian question for Islamic astronomy, by asking why Muslim scientists did not take the leap forward and initiate the scientific revolution even earlier than Europe.

The first striking difference that we see in Sayılı, compared to Adıvar and earlier reflections on the history of Islamic science in Turkey, was the addition of the concept of " scientific revolution " in his writings. Sayılı was convinced that World history had its most crucial turning point with the scientific revolution, which was around the time of Ottoman siege of Vienna. He once wrote that even if Vienna had fallen to the Turkish commanders, this event could hardly have changed the general situation. " The Ottoman Empire had fallen behind Europe ; she had already entered upon a period of decline, especially when compared with the West... Newton had finally published his monumental work, the Principia, in 1687, only four years after the Turkish siege of Vienna, while Ottoman men of learning were not as yet quite ready even to give Copernicus the credit that was due to him ; nay, they were not even duly aware of this great work, nor of those forming its more immediate sequels, such as the masterworks of Galilee and Kepler "[12].

Furthermore, for Sayılı, like many other intellectuals in his generation, Ottoman society was not only isolated from Europe and therefore missed the chance of learning from them, it also went into a process of decline in intellectual vitality. In this way the initial gap between the world of Europe and Ottoman society became bigger[13]. But Sayılı's idea of decline also imply that Ottomans or Muslims themselves lost the chance of initiating a scientific revolution due to their own decline, rather than simply in comparison to Europe. Sayılı's emphasis on the inability of Ottoman scientific community to create a scientific revolution in astronomy reflects the importance he gave to the significance of scientific revolution in world history. The idea of the Scientific Revolution, which had its origins in the 1920s but established itself in social science literature in 1940s, was not a significant theme in Adıvar's writings, though we have a similar emphasis on the groundbreaking radical changes in the scientific developments in Europe. When Sayılı was writing on Islamic and Ottoman science, the concept of " Scientific Revolution " was already elabo-

---

11. For a work of one of Sayılı's few students who continued the work in a similar direction, see Sevim Tekeli, *The Clocks in Ottoman Empire in the 16th Century and Taqi al Din's " The Brightest Stars for the Construction of the Mechanical Clocks "*, Ankara 1966.

12. Aydın Sayılı, " The Place of Science ", 26.

13. *Idem*, 28.

rated in influential works of Koyré, Burtt and Butterfield[14]. The above quotation from Sayılı, on the comparative significance of Vienna defeat and Scientific Revolution, seems to share the convictions of Herbert Butterfield's famous phrase. " (The scientific revolution) outshines everything since the rise of Christianity and reduces the Renaissance and Reformation to the rank of mere episodes, mere internal displacements, within the system of medieval Christendom "[15]. It should be remembered that this extraordinary emphasis on the scientific revolution in explaining the " rise of the West " or " the emergence of modernity " became dominant only after 1930s, even though the scientific flourishing in Europe were always part of the story of Western civilization. Recently, Floris Cohen and Toby Huff's studies on the emergence of modern science and " Scientific Revolution " also took scientific revolution as the key transformative event of European modernity[16]. Moreover, in both Cohen and Huff, the contingent association between European history, culture, society and the modern science has been further associated with liberalism, freedom and rationality, not simply with power as it was the case in Butterfield and Sayılı.

Aydın Sayılı initially shied away from one or another essential character ascribed to Islamic civilization as an explanation for the final rise of European astronomy and decline of Ottoman astronomy with careful contextual comparative studies. However, at the final analysis, Sayılı also subscribed to a culturalist explanation of the Ottoman decline in science, by taking the " faith " as the prime explanatory unit. He argued that a failed reconciliation between faith and science in Islamic civilization led to the decreasing dynamism of science, thus consequently to the non-emergence of modern science. There might be several factors that encourage Sayılı to an essentialist faith model rather than a sociological one. For example, he regards cultural and intellectual causes more directly tractable than social, political and economic factors. Furthermore, he uses a culturalist explanation by attributing to the " faith " a prime explanatory quality for understanding the developments in science in both medieval Islamic and European civilization, because of the power of religion in both of these societies at that time. Since both societies were at bottom theocratic, as he assumes, the crucial challenge for their success in science was to find " a solu-

14. Koyré's first book on scientific revolution, *Études Galiléennes*, appeared in 1939. For a summary of his ideas in English, see, Alexander Koyré, *Metaphysics and Measurement : Essays in Scientific Revolution*, M. Hoskin (ed.), Cambridge, Harvard Un. Press, 1968 ; E.A. Burtt, *The Metaphysical Foundations of Modern Physical Science : A Historical and Critical Essay*, London, Routledge and Kegan Paul, 1972 (reprint of the 1932 edition) ; H. Butterfield, *The Origins of Modern Science, 1300-1800*, revised edition, London, Bell, 1957 (1st ed., 1949).

15. Herbert Butterfield, *The Origins of Modern Science, 1300-1800, op. cit.*, vii. See also esp. Chapter 10 : " The place of scientific revolution in the history of Western Civilization ".

16. Toby E. Huff, *The Rise of Early Modern Science : Islam, China, and the West*, Cambridge University Press, 1993. Floris Cohen, *The Scientific Revolution : A Historiographical Inquiry*, Chicago, University of Chicago Press, 1994.

tion of the relations of faith and knowledge in a manner favorable to science and philosophy. This was, for medieval thinkers, a major intellectual task on which the fate of medieval science depended in great measure "[17]. However, Sayılı did not provide any detailed evidence and historical proof that faith precluded scientific dynamism, as there was no major debate on science and faith in the Ottoman time. One can understand that as an intellectual historian, Sayılı feels more comfortable with cultural analysis, yet, the absence of evidence in his argument for failed reconciliation between science and faith among the Ottoman learned community is a very striking shortcoming of his thesis[18].

Aydın Sayılı applied the primacy of the faith paradigm for the failure of Tanzimat reforms as well. After he noted the general difficulty of catching up with Europe in the non-European world during the 19th century, Sayılı repeated previous culturalist arguments to explain Ottoman "failure" in reforms, by noting that "there were the retarding forces of fanaticism and the inertia of established tradition". As historical evidence, he referred to the limitations on the printing press by the Ottoman authorities that specified that printing press have to be for the purpose of printing books not pertaining to religion and jurisprudence[19]. Interestingly, elsewhere in the same article, Sayılı notes that during the rebellion that ended the Tulip Period the printing press introduced from Europe in 1729 remained intact, which he interprets as being due to the formal sanction of religious authority for the printing press[20]. Contrary to what Sayılı asserts, it is historically more plausible to argue that limitation of the scope of the printing press to the non-religious sciences might had had a positive impact on the development of sciences rather than a negative impact, since it get rid of the competition of the religious books for the printing press and left publications in secular sciences free.

In short, in spite of variations in the explanations by Adnan Adıvar and Aydın Sayılı, there is an agreement that the root cause of the decline of Ottoman science is to be found in the Islamic faith, and in the ability of its orthodox upholders to stifle once-flowering science. Works by Adıvar and Sayılı represent the most refined and influential versions of Republican Islam-negative culturalism, which was carried over from the legacy of the Young Turk intellectuals and shared by official positivist view of historical evolution from Ottoman past to modern Republic. Many of the themes of this culturalism were present in the high school textbooks provided by Turkish Ministry of Education, although its various details changed over time. And the same arguments

17. A. Sayılı, *Observatory in Islam, op. cit.*, 410.
18. *Idem*, 411-412.
19. *Idem*, 48.
20. *Idem*, 39.

was repeated by the intellectuals of the Left, with the only possible difference that Islamic fanaticism and dogmatism were interpreted as an expression of certain backward feudal interests.

" Islam-Positive " Culturalist Historiography

Islam-negative culturalism was not the only historical narrative that the Turkish intellectuals had with regard to Islamic-Ottoman history of science. There has been another intellectual tradition that extends from Namık Kemal to Mehmet Akif Ersoy and Bediuzzaman Said Nursi that argued for the essential compatibility between Islamic faith and the modern science, and perceived history from this perspective. For instance Said Nursi was a prolific intellectual and religious leader who popularized " Islam-positive culturalism ". As Ibrahim Kalın showed, for Muslim modernists such as Said Nursi, " modern science is nothing but Islamic science shipped back to the Islamic world via the ports of European Renaissance and Enlightenment "[21]. Said Nursi not only claimed the compatibility between Islamic faith and modern science, but also appropriated modern scientific findings as confirmations of the Islamic faith, especially revelations. Ironically, Said Nursi had a scientistic world view as strong as Islam-negative culturalists such as Adıvar and Sayılı, who perceived secular and positivist science as the triumph of reason over against religion. However, scientism of Nursi and his followers' perceived and interpreted modern science as an instrument and confirmation of the theistic conception of universe. Politically, critiques of the Republican positivism through a synthesis of scientism and Islamic texts aimed to take away the authority and legitimacy of secular elites, who initiated a cultural revolution in the name of science and progress. Two competing interpretive traditions of Islam-based culturalism were also symbolizing a wider cultural-ideological conflict in Republican history between secular positivism and Islamic modernism[22].

Both Islam-negative and Islam-positive culturalism had major contradictions. The former usually has a difficulty in explaining the achievements of Islamic science in its history, especially when religious zeal was also very strong. For example, in most of the cases, they had to disregard the fact that Averroes was also a pious orthodox jurist, and Takiyüddin and Katip Çelebi, two Ottoman scholars praised as exceptions to the Ottoman decline in science, were no less religious than other Ottoman Muslims. The " orthodox faith vs. science " paradigm seemed so weak that, the actual history of Islamic science

21. Ibrahim Kalın, " Three Views of Science in the Islamic World ", in ed. Ted Peters, M. Iqbal and S. N. Haq (eds), *God, Life and the Cosmos : Christian and Islamic Perspectives*, , Ashgate Publishing Ltd., 2002 (Forthcoming).

22. Şerif Mardin, *Religion and Social Change in Turkey : The Case of Bediuzzaman Said Nursi*, New York, SUNY Press, 1989.

that it relies on can be used as to induce and sustain an opposite model, as developed by Islamic modernism and salafism, arguing that it was the pious and practicing Muslims who showed great achievements in the history of Islamic or Ottoman science. On the other hand, Islam-positive culturalism had a difficulty in explaining the decline of scientific activity among Muslims compared to Europe, given their thesis that Islamic texts give inspiration to modern scientific development. For example, Said Nursi or other Muslim modernists evade the issue of the failure of Ottoman scholars to catch up with the European scientific development. In fact, Islam-positive culturalists rarely deal with the Ottoman science, because of their belief in the golden age of Islamic science until the 13th century, and its subsequent decline parallel to the deviation from the original faith during the medieval period. Islam-positive culturalists in Turkey, who are usually proud of Ottoman past achievements in political organization and military conquest, still blame Ottoman educational institutions for not producing scientific progress. Ironically, both of the opposing culturalist models are based on the paradigm of " primacy of faith " to explain the various developments in the history of Islamic civilization, and they concur that Ottoman science after 17th century is not worth exploring.

Given the above-mentioned ideological context of the interest in history of science in Turkey, it is no coincidence that the writings of Seyyed Hossein Nasr on Islamic science were all translated into Turkish and found a very receptive audience among the circles sympathetic to Islam-positive culturalist explanations, but willing to overcome the scientism and positivism of the earlier generations. Hossein Nasr was among the early recipients of Ph.D. degree in Harvard's History of Science Department and thus has a shared educational background with Aydın Sayılı. He developed an alternative view of history of Islamic science within the Western academia by utilizing both metaphysical and epistemological critiques of modern science[23]. Thus, Hossein Nasr did not share the scientistic approach of Said Nusri and previous Muslim modernists who took modern science as objective truth and highly compatible with Islamic ideals. On the contrary, in a non-apologetic and confident style, Hossein Nasr argued that Islamic science is not only different from the modern Western science due to its spiritual dimension but also morally superior to it because of its harmony with nature and the Divine. Nasr's metaphysical critiques of modern science, some argued, made Nasr to write a biased account of history of Islamic science, in the sense that it always emphasize the Gnostic

23. Almost all the works of Seyyed Hossein Nasr have been translated into Turkish language during the 1980s and the 1990s. For the most influential work, see Seyyid Huseyin Nasr, *Islam'da Bilim ve Medeniyet*, Istanbul, Insan Yayınları, 1983. For its English version, see Seyyed Hossein Nasr, *Science and Civilization in Islam*, Barnes and Noble Books, 1992 (originally published in 1968). Another best-seller of Hossein Nasr in Turkey was the translation of, Seyyed Hossein Nasr, *Islamic Science : An Illustrated Study*, Kent, World of Islam Festival Publishing Company Ltd., 1976.

and philosophical aspects of Islamic science — with the purpose of clarifying how it can be an alternative to the current quantitative science.

Nasr' interpretation of the history of Islamic science was not harmonious with the legacy of Islam-positive culturalist view of history of science in Turkey. According to Nasr, modern science was not a continuation of the Islamic science, because it is based on a different metaphysical view of nature and men[24]. However, translations of Nasr' works into Turkish influenced the level of confidence of Islam-positive culturalism, as it usually dealt with the " golden age " of Islamic civilization. Nasr' writings on history of science did not deal with the arguments that forms the real strength of " Islam-negative culturalism " in Turkey, namely explanation of the decline of the Ottoman-Islamic science, and the failure to catch up with the modern science.

Ekmeleddin Ihsanoğlu's Critique of the Culturalist Approaches

Professor Ekmeleddin Ihsanoğlu, the founder of Istanbul University's History of Science Department and currently President of IUHPS/DHS, revised the predominant culturalist historiography with an explicit sociological agenda since the beginning of the 80s. Initially, he began two parallel projects. The first was an attempt to expand our historical knowledge on the scientific manuscripts and scientists of the Ottoman era. This research project culminated in the groundbreaking editions of catalogues on Ottoman mathematicians, geographers and astronomers[25]. Detection, cataloguing, and assessment of the Ottoman scientific materials have been overlooked due to previously predominant conviction that Ottomans failed in scientific field, and thus their writings do not have any contemporary value for study. Long years of primary source manuscript evaluations in various projects led by Ekmeleddin Ihsanoğlu demonstrated that Ottoman history, during the long period of its history, included a vibrant community of scholars that produced rich and diverse body of scientific works. Unearthing and simply describing the contents of Ottoman scientific literature gave the necessary database for the sociological study of the Ottoman science that can go beyond the duality of Islam-positive and Islam-

24. For the best analysis of Seyyed Hossein Nasr' view of sacred science and Islamic science, see Ibrahim Kalın, " The Sacred versus the Secular : Nasr on Science ", in Lewis Edwin Hahn, R. Auxier and L. Stone (eds), *The Philosophy of Seyyed Hossein Nasr*, Chicago, Open Court, 2001, 445-468.

25. For the three major collections that contains biographical data of Ottoman scientists and science manuscripts in the fields of geography, astronomy and mathematics, See Ekmeleddin Ihsanoğlu (ed.), *Osmanlı Astronomi Literatürü Tarihi* (History of Astronomy Literature during the Ottoman Period), Istanbul, IRCICA, 1997 ; *Osmanlı Matematik Literatürü Tarihi* (History of Mathematical Literature during the Ottoman Period), Istanbul, IRCICA, 1999 ; *Osmanlı Coğrafya Literatürü Tarihi* (History of Geographical Literature during the Ottoman Period), Istanbul, IRCICA, 2000.

negative culturalist generalizations. Similar primary source gathering and analysis by colleagues and students of Ihsanoğlu at the History of Science Department of Istanbul University created the long-neglected field of History of Ottoman Science within the larger framework of history of science and technology in Islamic civilization[26].

Ihsanoğlu interpreted conclusions of new researches and reassessed the previous arguments of both Islam-negative and Islam-positive culturalism. First, it became clear that, Ottoman scientific and intellectual life was not stagnant, as the rich literature on mathematics, geography and astronomy contains evidence for a continuing interest and scholarly activity on these subjects. One of the reasons why previous historiography overlooked the rich collection of scientific manuscripts was the constant comparison with the scientific developments in Europe. However, when evaluated in its own historical trajectory without making judgments based on European comparisons, there was not any decline in the Ottoman scientific activity. Second, disagreeing with Adnan Adıvar, Ekmeleddin Ihsanoğlu argued that Ottoman intellectuals were largely aware of the scientific and technical developments in Europe since the 17th century. In fact, thanks to frequent interaction with Europe, Ottomans transfered science and technology from Europe according to their needs and interest. For example, military technology was well followed, while astronomical innovations were easily accepted. Third, contrary to what is assumed in the Islam-negative culturalism, Ihsanoğlu's research showed that there was not any apparent conflict between science and Islamic faith both during the classical Ottoman period and the period of the transfer of science from Europe. The positive embrace of the Copernican astronomy and other scientific developments indicates that Ottoman Muslims did not experience anything similar to the controversies over the new astronomy in Europe. Ihsanoğlu further problematized the issue of science versus religion in the Ottoman appropriation of the Western science since the 17th century, and argued that the attitude of the Ottoman ulema and scientific community toward the European science was overall positive[27].

---

26. For more information on the research at the History of Science Department of Istanbul University, see Feza Günergün (ed.), *Osmanlı Bilimi Araştırmaları* (Studies on Ottoman Science), Istanbul, Edebiyat Fakültesi Yayınları, 1995. A dissertation by a student of Ihsanoğlu, the late Cevat Izgi, was a groundbreaking work on the education of sciences in Ottoman educational institutions. See, Cevat Izgi, *Osmanlı Medreselerinde Ilim*, Istanbul, Iz Yayınları, 1997.

27. See Ekmeleddin Ihsanoğlu, " Some Critical Notes on the Introduction of Modern Sciences to the Ottoman State and the Relation Between Science and Religion up to the End of 19th Century ", in eds. J.L. Bacqué-Grammont and E. van Donzel (eds), *Comité International d'Études Pré-Ottomanes et Ottomanes* : VIth Symposium, Cambridge, 1st-4th July 1984, Paris, Leiden, Istanbul, The Divit Press, 1987, and " Introduction of Western Science to the Ottoman World : A Case Study of Modern Astronomy (1660-1860) ", in E. Ihsanoğlu (ed.), *Transfer of Modern Science and Technology to the Muslim World*, Istanbul, IRCICA, 1992.

As for the question of relative failure of Ottoman reforms in scientific institutions and technological transformation, Ekmeleddin Ihsanoğlu attributed its causes to social and political reasons, rather than simply cultural ones or to mentality and religion. He rejected the argument that Ottoman ulema was against the Western science and technology, and furthermore, he claimed that they were rather overall supportive of the transfer of science and technology from Europe. In the cases where there emerged a conflict between ulema and the new type westernized intellectuals — which was taken as an indication of the eternal conflict between science and religion —, Ihsanoğlu maintains that it was due to social context of Ottoman reforms. For example, the graduates of new Western style schools were privileged in terms of their bureaucratic position and salaries, causing a certain grudge among the *madrasa* graduates[28]. To explain the deeper causes of Ottoman failure to catch up with the scientific progress in Europe, Ihsanoğlu underlines the functionalist conception of science and economy of the technological improvements, not religious dogmatism, which made Ottomans just keep up with the practical requirements of the age rather than appropriating new scientific paradigms in Europe[29].

Finally, Ekmeleddin Ihsanoğlu's revisions regarding the history of Ottoman science led him to deal with the question of historiography, as he had to explain why and how an essentialist culturalism dominated the interpretation of previous studies in this field. In a recent research project on the history of Ottoman *madrasa*, Ihsanoğlu devoted his first article purely to the historiography, exploring how the late Ottoman era debates on *madrasa* reforms and the Republican revolutions in the field of education shaped an ideological perspective in history writing on the subject. As one group of scholars created a historical narrative that Ottoman educational institutions were already obsolete and had little potential for revival and reform according to the requirements of the modern age, another group of scholars exaggerated the role of natural sciences in Ottoman *madrasa* in order to argue for its capability to adapt to modernity. In the end, we have a somehow distorted narrative of the Ottoman *madrasa*. Any new research on Ottoman science and educational institutions has to take ideological context of the previous literature and must rethink the secondary literature to avoid the modern myths of positivist historiography[30].

The Turkish intellectual circles have welcomed the new sociological approach to the history of Ottoman science. The reception of the segments of

28. E. Ihsanoğlu, " Some Critical Notes on the Introduction of Modern Sciences to the Ottoman State and the Relation Between Science and Religion up to the End of 19th Century ", *op. cit.*

29. E. Ihsanoğlu, " Ottomans and European Science ", in P. Petitjean (ed.), *Proceedings of the International Colloquium 'Sciences and Empires'*, Netherlands, Kluwer Academic Publisher, 1992, 37-48.

30. E. Ihsanoğlu, " The Initial Stage of the Historiography of Ottoman Medreses (1916-1965). The Era of Discovery and Construction ", *Archivum Ottomanicum*, 18 (2000), 41-85.

Turkish reading public traditionally associated with the left and positivism is exemplified by the publication of Ekmeleddin Ihsanoğlu's articles by the leading new-left publishing house, İletişim Yayınları[31]. The publication and success of this book indicates a change in the approach of circles usually inclined towards Islam-negative culturalism in the direction of a better appreciation of the complexity of intellectual life during the Ottoman times. Meanwhile, intellectual circles sympathetic to the Islam-positive culturalism too welcomed the writings of Ihsanoğlu, as it offered a sociological critique of the historiography that blamed Islamic beliefs for the backwardness of the Ottoman science. However, the same sociological approach methodologically challenges the contradictions of Islam-positive culturalism that attributes the " golden age " of Islamic science (from 9th to 12th centuries) to the impact of Islamic worldview, but dissociates Islamic belief from the " relative " decline of science in the Muslim societies.

CONCLUSION

Rethinking the Role of Culture in the New Agenda
for the History of Ottoman Science

The recent turn to social history of Ottoman science does not mean that culture and mentality, or Islamic tradition did not matter in influencing the scientific activities at all. The challenge for the post-Ihsanoğlu research agenda for Ottoman science is to re-examine the role of culture, which is not simply reducible to Islamic faith, on the nature of scientific activity through a sociology of knowledge. Sociological approach acknowledges that culture still matters because, as a set of socially significant values, it affects the behavior and activities of the historical actors. The mistake of the essentialist arguments of culturalism was to assume a monolithic Islamic faith without really specifying its historical context, social meaning and practical impact.

In this point, an application of the concept of " scientific paradigms ", together with a more sociological examination, can teach us how Islamic tradition mattered in Ottoman encounter and appropriation of the modern science. Initially, the ideological nature of the historiography on science in Turkey precluded the application of Thomas Kuhn's concept of paradigm to the encounter of Ottoman scientific community with the West[32]. Aydın Sayılı would have personally known Thomas Kuhn and was familiar with his argument, but never addressed the question of the scientific paradigms in his works on the absence of scientific revolution in Islamic astronomy. Since we are beginning to have a

---

31. E. Ihsanoğlu, *Büyük Cihad'dan Frenk Fodulluğuna*, Istanbul, Iletisim, 1996.
32. T. Kuhn, *The Structure of Scientific Revolutions*, 3rd ed., Chicago, University of Chicago Press, 1996.

better picture of the nature of the Ottoman scientific community and their encounter with the emerging modern science in Europe, we can now explore if their scientific paradigms differed, and whether this difference had any role in shaping the peculiar Ottoman response to the European science. Once we approach the topic beyond the question of who to blame for the later " backwardness " of Ottomans in science and technology compared to Europe, new studies on the encounter of Ottoman scientific tradition with the emerging modern science could help us to reassess not only the characteristic of Islamic science as it developed during the Ottoman era, but also the nature of modern science itself through the mirror of its reception by the Ottoman scientific community from the 17th to the 19th centuries.

In short, a survey of historiography on Ottoman science shows that ideological context of the Republican era and the culturalist assumptions about the relationship between Islamic faith and scientific activity precluded a contextualized, nuanced history of the Islamic-Ottoman history of science. However, the methodological and historiographic critiques of this inherited culturalist polarization, as well as monumental data collection and sociological analysis in the works of Ekmelleddin Ihsanoğlu and his colleagues created a new research agenda for Ottoman history of science. This new research agenda will not only help ease the ideological polarization in Turkish intellectual life through overcoming the narrow views of both Islam-positive and Islam-negative culturalism, but also offer a new perspective on the relationship between cultural diversity and transnational history of science through the case study of the encounter of Ottoman scientific tradition with the modern European science.

THE STATE AND PROFESSIONAL IDENTITIES : THE EMERGENCE OF THE SOCIO-PROFESSIONAL CLASS OF GREEK ENGINEERS AT THE BEGINNING OF THE 20th CENTURY

Yiannis Antoniou and Michalis Assimakopoulos

This paper constitutes our attempt to describe the circumstances that encouraged the configuration and development of the engineering profession in Greece within the first forty years of 20th century.

The emergence of this profession provides us with a double-sided portrait. Until the 1880s, the initial portrait displayed the limited role in Greek society played by the modern technical professions, which continued to employ traditional techniques for producing most artifacts. Few persons in engineering held any academic qualifications and the profession lacked any institutional consolidation. For instance, in 1900 there were 149 qualified engineers, who were graduates from the Polytechnic School of Athens and according to the registry of the *Technical Year Book of Greece*, 1934, 66 graduates from European technical schools[1, 2].

1. These 66 graduates from European technical schools represent only a part of the engineers who graduated the last twenty years of 19th century and they were still alive in 1934. The real number is bigger but there are not still precise rundown elements for them. For example, 23 Greek engineers graduated from Federal Polytechnic School of Zurich until 1900, Alexandros Tsatsos, *Epetiris ton Phoitisanton Ellinon eis to Omospondiakon Polytechnion tis Zirihis, 1855-1978* (*Year Book of Greek Students in Federal Polytechnic School of Zurich, 1855-1978*), Athens, 1978, 84. Additionally, during the 19th century, about 180 Greek students studied in French *Grandes Écoles*, F. Assimakopoulou, C. Chatzis, " Les élèves grecs dans les Grandes Écoles d'ingénieurs en France au XIXe siècle " in the same volume.

See also : *Techniki Epetiris tis Ellados* (*Technical Year Book of Greece*), vol. Β', Athens, Technical Chamber of Greece, 1934 ; Agellos Ginis, *Peri tou Scholeiou ton Viomihanon Technon* (*About the School of Industrial Arts*), Athens, 1912, 5 ; *Odigos Spoudon Ε.Μ.Π., 1950* (*Guide for Studies in the Polytechnic School, 1950*), Athens, National Technical University of Athens, 1950, 174 ; Costas Biris, *I Istoria tou Ε.Μ.Π.* (*History of the Polytechnic School*), Athens, National Technical University of Athens, 1957, 328.

2. The function of Greek engineers within the broad framework of the Ottoman Empire is a valid field of study of its own right.

During the second decade of the 20th century, this picture changed drastically, including, at this time, an integrated network of professional, educational, administrative and entrepreneurial institutions, which impetuously ushered in the admission of a new socio-professional class that aspired to dominate the social space. In 1934, there were 2338 qualified engineers : 664 of these graduated from foreign technical schools, while the remaining 1674 graduated from the Polytechnic School of Athens. These 2338 engineers were divided into the main specifications of engineering : 1015 civil engineers, 245 architects, 309 mechanical engineers, 354 electrical engineers, 133 chemical engineers, 41 mining engineers, 17 naval architects, and 224 surveying engineers. Furthermore, these professionals were organized within a powerful official institution, the Technical Chamber of Greece[3]. During this period, the Polytechnic School of Athens (National Technical University) monopolized the organic reproduction of the profession, decisively securing, in the process, the scientific status of its graduates. Additionally, 97 technical companies, most of them established after 1920, defined their constructive, commercial, and industrial activity with the precipitous diffusion of engineers into the business sector[4].

The hypothesis that we wish to test in this paper concerns the two factors that encouraged the change for the emergence and for a certain formation of the profession. The first factor relates to the configuration and development of the Greek State and its intention to establish itself within the modern era. The second factor, directly related to the first, concerns the promotion of an ideology that tended to identify the concept of progress with that of the development of science and technology[5]. For the second, the Greek engineers tended to introduce themselves as the exclusive representatives of this ideology.

Peter Meiskins and Chris Smith adopt an analytical category that explicates four models relevant to the configuration of engineering profession internationally, namely, the " craft ", " managerial ", " estate ", and the " company-centered " models[6]. The first model parallels the professional configuration in Britain, the second case for the U.S.A., the third in France, and the fourth in Japan. Within Germany, however, a combination of the second and the third cases prevails.

---

3. *Techniki Epetiris tis Ellados* (*Technical Year Book of Greece*), vol. B' (1934), *op. cit.*

4. *Ibid.*

5. Yiannis Antoniou analyzes this theme within his PhD. Dissertation (Department of Humanities, National Technical University of Athens).

6. Peter Meiskins & Chris Smith, *Engineering Labour, Technical Workers in Comparative Perspective*, London , New York , Verso, 1996, 233-255.

In the first model, apprenticeship during the production procedure shapes the engineering profession. The individual assessment of the engineer (and the established hierarchies of the profession) does not depend on academic or educational qualifications, but they are consequences mainly of a shop floor culture. In the second case, the configuration of the profession depends upon the development of academic educational institutions, and this managerial model is directly associated with the form of administration and organization for the enterprises. In this case, engineering becomes identified with the management of men and machinery. Next, the configuration of the profession that depends upon the stratified development within the technical professions defines the third case. Technical education, itself stratified, bestows the criterion for this professional delineation, which the state officially guarantees. According to this model, qualified engineers occupy the pinnacle of this pyramid and identify the essence of their profession with this position, in essence dissociating themselves from management and other technical professions. Finally, in the fourth case, the rules of organization and operation of the specific enterprise configure the profession. According to this model, academic qualifications are the prerequisite for entry into engineering, but the enterprise determines the constitution of the company hierarchy.

Within the following passages, we will support our hypothesis by identifying the corresponding concepts regarding this taxonomy with the conditions that encouraged the emergence and the arrangement of the engineering profession in Greece.

Greek engineers are both the products of and the producers for a peculiar institutional and technological innovation[7]. We can locate the start of this change within the institutional separation of learning procedures for various technical skills from the professional practice of these skills, that is, the separation of technical education from relevant work experience[8]. This innovation, apart from its institutional and technological parameters, also included an epistemological one counterpart. Namely, it introduced a perspective of the world based upon scientific abstraction of objects and, subsequently, reinforced a theoretical apprehension of reality. This perspective led to the understatement of

7. Aliki Vaxevanoglou, *I koinoniki ipodohi tis kainotomias, to paradigma tou exilektrismou stin Ellada tou Mesopolemou* (*The Social Reception of the Innovation, the Paradigm of the Electrification in Inter-War Greece*), Athens, National Hellenic Research Foundation, 1996, 84.

8. This separation, although established around 1830 in the Military Academy, occurred in the Polytechnic School during the 1860s. See Y. Antoniou and M. Assimacopoulos, " The Military Academy and the School of Arts in 19th Century Athens : Two Types of Engineering Education ", which will be published within the proceedings of the 2000 conference held in Syros, " Science and Technology in 19th c. : the role of the army ".

the empirical model for the perception of reality, which is the defining feature of traditional techniques practiced by skilled craftsmen. With the establishment of the Modern Greek State in 1832, these modernized practices were imported into the country.

The establishment of the Military Academy of Greece — the *Scholi Evelpidon*, in 1828, following the model of the *École Polytechnique*, consolidated this educational innovation during the 19th century by promoting the status of military engineers[9]. Later, in 1836, with the establishment of the Polytechnic School of Athens, which initially operated as middle-level technical school adhering to the French model of the *Art et Métiers*, also promoted this transformation. In 1887, the Polytechnic School became an advanced technical school, this time following the German models for their Higher Technological Institutes. By this time, the institution included two separate schools : the School for Civil Engineers and the School for Machinists. The educational courses taught within the curriculum included mathematical analysis, physics, chemistry, and additional specialized technical lessons. The graduates from these schools were considered the equals with their counterparts from the European polytechnic schools on the educational and professional levels. During its 25-year history (1887 to 1914), these two schools produced a total of 321 engineers : 272 civil engineers and 49 machinists. This disparity between machinists and civil engineers correlated closely with the particular difficulties of establishing industrial development in Greece.

In 1914, the Polytechnic School became the academic equal to the University of Athens. According to the new regulations, which were applied gradually until 1917, five separate schools were established. The School of Civil Engineering, the School of Mechanical and Electrical Engineering, the School of Chemical Engineering, the Architecture School, and the School of Survey Engineering. The relevant codes provided for the establishment of four middle-level technical schools that fell under the supervision of the academic schools. Studies lasted for a duration of five years for the students within civil engineering, mechanical and electrical engineering, four years for the chemical engineering students and the architects, and three years for the surveyors and the students within the middle-level programs. These new incorporation laws also provided the Polytechnic School with the exclusive privilege to administer aca-

---

9. Epaminondas Stasinopoulos, *I Istoria tis Scholis ton Evelpidon, 1828-1953* (*The History of the Greek Military School*), Athens, 1933 ; Maria Kardamitsi-Adami, " Oi Protoi Ellines Mihanikoi " (" The First Greek Engineers "), *Technica Chronica* A, Athens, Technical Chamber of Greece, vol. 8, s. 4 (1988), 63-83 ; Andreas Kastanis, *I Stratiotiki Scholi ton Evelpidon kata ta prota xronia tis leitourgias tis, 1828-1834* (*The Military School during the First Years of its Operation, 1828-1834*), Athens, Ellinika Grammata, 2000.

demic technical education in Greece. During the period of 1915-1940, 1873 qualified engineers graduated from the Polytechnic School[10].

One can also trace the development of this change with the intervention of French technical missions during the 1880s, the implementation of the French administrative system in state services, and with the establishment of the Corps of Greek Civil Engineers[11]. These reformations carved their results into the modern configuration of the Greek urban space, especially concerning to the widening interference of the state within the public sphere. This interference, whether direct or indirect, was shaped largely through large technical constructions : roads, ports, shipyards, railways, drainage constructions, water reservoirs for Athens, new industries, electrification projects, urban transportation developments, communications projects, feasibility studies for the exploitation of mineral and water sources, and notably within 20th century, large scale land reclamation projects. These projects characteristically disfavored the traditional techniques craftsmen practiced. These projects also presupposed the development of a complex system of design, control, and labor organization. All these factors extended and homogenized the emerging class of Greek engineers. The establishment of the Ministry of Transportation in 1914 proved to be pivotal for improvement of these systems, and, not surprisingly, the first minister was an engineer. At the commencement of operations, this institution employed 50 engineers and 50 foremen, but by 1934, the number increased to 300 engineers and 300 foremen[12].

Through promoting a stratified system of technical education, this educational innovation was also legitimized ideologically. The pyramid configuration of educational institutions started by the final decade of 19th century and took its final shape during the inter-war period, coinciding with the establish-

10. " Register n° 10 ", *Archives of National Technical University*, " Tables of marker 1864-1865, 1880-1881, 1881-1890, 1890-1910 and 1910-1918 ", *ibid.*, " Reports of the meetings of the Faculty Council of the NTUA, 1909-1914, 1914-1917, 1917-1920 ", *ibid.*
  Techniki Epetiris tis Ellados (Technical Year Book of Greece), vol. B' (1934), *op. cit.* ; *Odigos Spoudon E.M.P. 1950 (Guide for Studies in the Polytechnic School, 1950)*, Athens, National Technical University of Athens, 1950, 174 ; C. Biris, *I Istoria tou E.M.P.*, *op. cit.*, 475 ; Eleni Calafati " To Ethniko Metsovio Plytechnio sto gyrisma tou eona. Epaggelmatikes diexodoi ton apofiton kai thesmiko kathestos tou idrymatos " (" The Polytechnic School to the turn of the century. Professional opportunities for the graduates and institutional status of the School "), *Panepistimio, Ideologies kai Pedia. Istoriki Diastasi kai prooptikes* (University, Ideologies and Education. Historical Perspectives), vol. A', Athens, Historical Archives of Greek Youth, 1989, 170 ; Ageliki Fenerli, " Spoudes kai spoudastes sto Polytechnio 1860-1870 " (" Studies and Students in Polytechnic School, 1860-1870 "), *ibid.*, 160-161.
11. Decree, XIIH, 1878, *Official Gazette*, n° 27, Athens 4/5/1878.
12. Agelos Economou, " Ta Dimosia Erga kai ai Dimosiai Technikai ipiresiai " (" The State Constructions and the State Technical Administration "), *Techniki Epetiris tis Ellados (Technical Year Book of Greece)*, vol. A' (1935), *op. cit.*, 213-278 ; Dimitrios Arliotis, " Ta idravlika erga en Elladi " (" The Hydraulic Constructions in Greece "), *ibid.*, 278-311.

ment and operation of many low and mid-level technical schools in Athens, among other cities. This hierarchy certified varying levels of accessibility for technical and scientific knowledge, produced new techniques, and established alternate social groups. These groups were assigned the task of putting these new techniques into practice by following a modern classification context, which changed the spectrum of technical professions in Greece. These institutions for the technical education — and the state-sponsored hierarchy — produced additional professional authorities and identities. These identities, especially for the engineers, proved to define the essence of their social self–consciousness[13].

Drawing from our researches, the emergence of the so-called " scientific industries " and " scientific industrialist ", during the first decade of the 20th century secured the marriage between engineers and the idea of scientific rationality. The chemical, cement, electrical, telecommunication, and the construction industries created a fresh entrepreneurial field and a radical type of entrepreneur, notably different from the accepted standards of the Greek version of capitalism[14]. In many cases, this new class of Greek entrepreneurs established affiliations with Greek engineers, principally with those who graduated from German-speaking polytechnic schools and especially with a six–person group — the " Circle of Zurich " — who graduated from the Federal Polytechnic School of Zurich[15]. These people virtually established the emergent industries, they decisively contributed to the social and institutional formation of the Greek industrialist class, and they eventually became the luminaries of Greek industrialization during the Inter-War period.

13. Stratis Papaioannou, " Ena Protoporiako idrima kai i epohi tou " (" The Age of an Innovative Institution "), *Economikai, Kinonikai Spoudai*, s. 1, Athens, Higher School of Industrial Studies, 1951, 241 ; Agellos Ginis, *Peri tou Scholeiou ton Viomihanon Technon (About the School of Industrial Arts)*, Athens, 1912, 1-12 ; Ioannis Hatsopoulos, " To Ehnikon Metsovion Polytechnion " (" The National Technical University of Athens ", *Techniki Epetiris tis Ellados (Technical Year Book of Greece)*, vol. A' (1935), *op. cit.*, 167-173.

14. Chr. Agriantoni, " Viomihania " (" Industry "), *Istoria tis Ellados ton 20o Aiona (History of Greece in 20th century)*, Athens, Vivliorama, 1999, p. 184.

15. The members of the " Circle " were Alexandros Zahariou, Nicolaos Kanellopoulos, Kleonimos Stilianidis, Leontios Economidis, Leonidas Arapidis and Andreas Hajikiriakos. In the *Technical Year Book of Greece* from 1934, the concise biographies of 56 engineers are recorded, all of who were graduates of the Federal Polytechnic School of Zurich. 35 of them graduated during 1890-1920. *Techniki Epetiris tis Ellados (Technical Year Book of Greece)*, vol. B' (1934), *op. cit.* ; Constantinos and Spiridon Vovolinis, *Mega Biographicon Lexicon (Biographical Dictionary)*, vol. A', Athens, Viomichaniki Epitheorissi, 1958, 46-57, 73-83, 85-102, 105-111, vol. B', 311-328 ; Eleni Calafati, " Alexandros Zahariou, o kat' exohin michanikos tis ellinikis viomihanias " (" Alexandros Zahariou : the Engineer of the Greek Industry "), " Vivliothiki kalliston anagkaiounton vivlion kai omologoumenos kalliston ephimeridon ", *I palies silloges tis vivliothikis tou E.M.P. (The Old Collection of the Polytechnic School Library...)*, Athens, National Technical University, 1995, 119-144.

The main challenge for these engineers, entrepreneurs, and industry executives was the creation of a " scientific " administration for technological innovation and promote their dominant position in productive, entrepreneurial, and social circles. They connected their domination and authority with a goal of restructuring the technological and industrial priorities for both the present and future development of Greece[16].

During the first twenty years of 20th century, the momentum of the profession was intensified by the emergence and the development of the first professional engineering institutions, which, in 1923, were gathered for the establishment of the Technical Chamber of Greece. During the Inter-War period, the Chamber claimed two privileges : first, to be the official technical consultant to the state, and second, to be the sole representative for the professional concerns of Greek engineers. The Chamber's claims were advanced indirectly by institutional adjustments that increased the opportunities for admission for qualified engineers to enter the higher offices of state administration, directly by the Chamber's participation in designing the technical policies for the state, by the installation of standards to all sectors of construction, and by the engineering presence within the new institution of the Senate, which was established by the constitution of first Greek Republic in 1927[17].

Beyond these, the Chamber developed a network of relations with relevant international professional organizations and participated in congresses and other international meetings. It also developed a publishing network, the core of which was the journal *Technical Review* (*Technika Chronica*). The Chamber's policy exercised authority to exert professional closure, to control and reduce the number of Polytechnic School graduates, to restrict the professional opportunities for foreign engineers working in Greece, to demarcate the professional protocols for several engineering projects, and to exclude craftsmen from officially recognized engineering professions. The antagonisms that arose between the qualified engineers and other professional groups — mainly

16. The engineers who belonged to this group (or who existed on its periphery) interfered drastically with the dialogue surrounding the public misgivings about the nation's industrial development and evolution. Also, through the Association of the Greek Industrialists and other entrepreneurial organizations, they tampered with social and cultural activities, while several persons held significant and influential political positions.

Indicative sources : C. and S. Vovolinis, *Mega Biographicon Lexicon,* vol. A', *op. cit.,* 46-57, 73-83, 85-102, 105-111, vol. B', 311-328. " Proto Panellinio Synedrio Georgias, Viomihanias kai Emporiou " (" First Greek Congress of Agriculture, Industry and Commerce "), D*eltion tis Emporikis kai Viomihanikis Acadimias* (*Journal of the Commercial and Industrial Academy*) (December 1909-January 1910), 111-123 ; Leontios Economidis, " Proodos tis en Elladi Chimikis viomihanias " (" Progress of the Greek Chemical Industry "), *ibid.* (1 April 1909).

17. The first engineering representative in the Senate was N. Kitsikis, a leading engineering figure during the inter-war period, professor of the Polytechnic School, and president of the Technical Chamber of Greece. *Techniki Epetiris tis Ellados* (*Technical Year Book of Greece*), vol. B' (1934), *op. cit.,* 147.

within construction and electrification — ended with major institutional victories for the Chamber and their qualified engineers[18].

Our thesis is that the initial institutional formation of the Greek engineering profession, its development during the Inter-War period, and its official ideology, although they appeared significant differences to their structure, resembled the French model. The exercise of its practice was connected with the possession of a position within an officially consolidated hierarchy of educational and professional institutions. Greek engineers — comprising state executives, independent entrepreneurs, or employees, especially under the jurisdiction of technical companies that entered the public works field and constructions at large — were identified socially with their profession and separated themselves from the social space of salaried work personnel.

The small group of industrial engineers and executives employed in these enterprises diverged somewhat from the above model. Instead, their profile resembled the German model, which combined the characteristics of the estate model with those of the managerial. In this latter case, engineers were defined by their exclusive capability to manage new technology, which was perceived as the exemplar for rational management of both enterprise and society. In any case, the professional field was constructed to have all the characteristics of a socio–professional elite. This construction related directly to state politics and was connected to the coordinated requisitions by the Greek professional engineering organizations. The rapid enlarging and gradual reinforcing of the professional field encouraged the development of engineering social prestige and the consolidation of their professional privileges.

We claim that the official ideology of the socio-professional class of the Greek engineers, which one can trace through all these evaluations, was stamped by the familiar ideology of technological determinism[19]. In this context, the technology was promoted as the prerequisite for the disavowal of Greece's oriental past and for the accomplishment of any modernist aspiration.

---

18. Indicative sources : *Techniki Epetiris tis Ellados* (*Technical Year Book of Greece*), vol. A' (1935), *op. cit.*, 8-52 ; Ilias Agelopoulos, " I egatastasi ton arhon tou TEE " (" The Settlement of the Administration of the Technical Chamber of Greece "), *Erga* (*Works*), 2 (1925), 44 ; I. Kandilis, " To zitima ton chimikon en Elladi, iperparagogi, anergia " (" The Problem of the Chemists in Greece, Overproduction and Unemployment "), *ibid.*, 3 (1925), 73 ; " To Diethnes Sinedrio tou technikou tipou " (" The International Congress of the Technical Press "), *ibid.*, 6 (1925), 153 ; " Kratos kai TEE " (" State and Technical Chamber "), *ibid.*, 10 (1925), 251 ; " To en Praga Sinedrion ton Epimelitirion Michanikon " (" The Congress of Engineering Chambers in Prague, March 9-12-1928 "), *ibid.*, 77 (1928), 137 ; " Peri tis chrisimopiiseos tou triphasikou revmatos " (" About the Utilization of the Three-phase Electric Power "), Proceedings of the meeting of the Electrical Engineers Department of the Technical Chamber of Greece, *ibid.*, 78 (1928), 159 ; " Report of the Technical Chamber's Administrative Committee March 19th 1934 ", *Technica Chronica* (*Technical Review*), 13 (1934), 305-335.

19. Merrit Roe Smith and Leo Marx, *Does Technology Drive History ? The Dilemma of Technological Determinism*, Cambridge, MIT Press, 1995.

This ideology also represented, in fact, an ideological alternative to the semi–official ideology of nationalism[20]. Greek engineers introduced themselves as the principle representatives for this orientation.

The modernization projects of Greece — conducted by Liberal Party and Eleftherios Venizelos, who dominated the Greek political life during 1910-1932, despite a few brief intermissions — became closely correlated with the configuration and reinforcement of the professional elite of engineers and the ideas that they promoted. On their part, the engineers defended and implemented modernization programs during this period, and they appeared to symbolically represent the ideals of modernization. Hence, they either demanded the acceleration and broadening of acceptable policies or condemned the forestalling and disintegration of their cherished modernist aspirations[21].

Their official intervention in public affairs seemed guided by a sovereign criterion, which submerged any conflict with their political choices, namely, the criterion for scientific rationalism and technological effectiveness. They promoted these goals not only as the essential elements of technological construction, but also as the constitutive conditions for the management of society as a whole. In this context, an individual engineer's synchronization with the general ideological profile of a political party (or political system) was not the first criterion for admission to politics. Politics itself submitted to this new ideological adjustment and the engineers proclaimed themselves its progenitors and guarantors. By analogy, the way in which technical journals saluted the election of the engineer Herbert Hoover to the American presidency in 1928, or their praise for the ideas of Henry Ford, F.W. Taylor and H. Fayol verify these assertions[22].

20. Leontios Ekonomidis, " I viomichania vasis tis ethnicis proodou " (" The Industry is the Base of National Progress "), newspaper *Embros*, 25/11/1918 and 4/12/1918, cited to C. and S. Vovolinis, *Mega Biographicon Lexicon,* vol. Β', *op. cit.,* 322.

21. Indicative sources : " Ta nea politica nefi na dialithosi to tahiteron " (" The New Political Clouds to be Dissolved Immediately "), *Erga (Works)*, 74 (1928), 53 ; *I economiki ereuna ton megalon techikon zitimaton (The Economical Inquiry of the Great Technical Issues)*, Athens, Techniko Epimelitirio tis Ellados (Technical Chamber of Greece), 1933. This is a massive volume that includes the papers from a forum, organized by Technical Chamber in 1932 and which lasted for five months.

22. Indicative sources : Leontios Economidis, " Proodi tis chimikis viomixanias en Elladi " (" The Progress of Chemical Industry in Greece "), *Archimides* (1st April 1909) ; " Herbert Hoover, the Miner Analogist President of the USA ", *Erga (Works)*, 83 (1928), 223 ; Kleanthis Filaretos, " To sidirometallevma entos 41 oron ginete aftokiniton. I organosi kai i egatastasis Ford " (" The Iron Ore in 41 Hours Becomes Automobile. The Settlements and the Organization of Ford "), *ibid.*, 27 (1926), 49 ; Th. Haritakis, " I epistimi tis organoseos tis ergasias " (" The Science of Labor Organization "), *ibid.*, 29 (1926), 97 ; " I didaskalia tis diikiseos ton ergasion " (" The Instruction of the Labor Administration "), *ibid.*, 30 (1926), 121 ; " O michanikos kai o viomixanos eine oi protergatai tou politismou mas " (" The Engineer and the Industrialist are the Creators of Our Civilization "), *ibid.*, 85 (1928), 379 ; " Oi megaloi michanikoi kai viomihanoi alla kai megaloi Americanoi dianooumenoi " (" The Great American Engineers and Industrialists are in the Same Time the Great Intellectuals "), *ibid.*, 90 (1929), 531 ; I. Varvagianis, " To sistima Taylor en ti viomihania " (" The Taylor System to the Industry "), *Viomihaniki Epitheorisi (Industrial Review)* (October 1916), 503-510.

When the ideological dilemma 'nation or society' erupted, the Greek engineers, according to their official ideology, chose 'society,' accepting as a definite fact the social diversification that accompanied technological development
and the expansion of industrial activity. At the same time, they tried to pacify
the fears from several social groups who considered industry as a threat to
social coherence. In doing so, they adopted the ideological principles of social
paternalism and contrasted it with the ideology of the class struggle, which the
communist party and the labor unions continually projected[23]. Falling into the
spirit of the times, the professional engineers connected their anticipation of
industrial development with the autarkic principles of economy and state pro-
tection for national industrial production[24]. Finally, they attempted to impose
the ideology of industrial development, correlated it with scientific rationality,
projected the marriage of science and industry as the prerequisite for the confrontation with underdevelopment, which they perceived as the principle
retarding agent for the impotent Greek economy.

The influence of these ideas can be traced to significant changes in the institutional and production fields, to the reinforced role of the engineers in this
emergent context, and also to the explosion of technical journals especially
within the Inter-War period[25]. These journals were the site into which a new
kind of public speech was born, called " technical literature ", following the
terminology of the age[26]. This site was also the fertile ground for the production of a modern type of intellectual, the " technical intellectual ". These men

23. In a meeting of the *Greek Chemists Association* on March 27, 1927, two groups conflicted
over the right of some members, who were industrialists, to participate within the administrative
committee of the Association. The group that supported the industrialists' participation accused the
opposition that the communists influenced their ideas (*Erga,* 44(1927)). In a subsequent issue of
the journal (*ibid.,* 46 (1927), 548), one member of the administrative committee defended the
group accused of bolshevism, alleging that the accusation was unfounded and the problem was the
result of a misunderstanding. In the same issue, K. Filaretos, a chemical engineer and director of
the journal, satisfied by these caveats and the propitious unfolding of the case wrote : " the theory
of the class struggle has not place in the community of scientists. All the chemical scientists in the
world are always colleagues and cooperators ". K. Zegelis, " I chimia paragon kinonikis
exisoseos " (" Chemistry as a Factor for the Social Equality "), *ibid.,* 87 (1929), 413 ; " Sintomos
anaskopisis tou ergatikou mas zitimatos " (" A Brief Review of our Labor Problem "), *ibid.,* 91
(1929), 557.

24. Indicative sources : Leontios Economidis, " Peri tis eghoriou viomihanias " (" About the
Domestic Industry "), newspaper *Acropolis,* 16/6/1918, cited to C. and S. Vovolinis, *Mega Bio-
graphicon Lexicon,* vol. B', *op. cit.,* 320-322 ; Leontios Economidis, " O polemos kai i anagi anap-
tixeos tis ethnicis oiconomias " (" The War and the Exigency of the Development of the National
Economy "), *Viomihaniki Epitheorisi* (*Industrial Review*) (1916), 417-461 ; A. Hajikiriakos,
Viomihaniki Politiki (*Industrial Policy*), Athens, Association of the Greek Industrialists, 1929 ;
Techniko Epimelitirio tis Ellados (Technical Chamber of Greece), *I Oiconomiki ereuna ton megalon techikon zitimaton* (*The Economical Inquiry of the Great Technical Issues*), Athens, Technical Chamber of Greece, 1933.

25. *Erga* (*Works*) and *Technika Chronica* (*Technical Journal*).

26. N. Kanellopoulos, a chemical engineer and the owner of the journal, emerged as one of the
celebrities within the inter-war Greek industry. He wrote that " the aim of the journal *Erga* is the
creation of Greek technical literature and the Greek technical writers ", *Erga* (*Works*), 87 (1929),
467.

confronted the intellectuals steeped in classical literature and challenged the ordering of the ideological and social hierarchies of 19th century Greece. Within this incipient ideological context, one could call the engineers, as the epitome of modernization, the organic (instrumental) intellectuals of the Greek bourgeoisie, using Antonio Gramsci's terminology[27].

This challenging ideology, in some versions, was connected with orthodox technocracy — transplanting Veblen's ideological schema to the Greek inter-war period —when legislative crises within the Parliament were at their peak[28]. Technocratic representatives among Greek engineers — especially A. Rousopoulos, an executive in the Technical Chamber and a professor in the Polytechnic —castigated the parliamentary system and promoted the authoritarian solution to the socio-political problems through the establishment of the " technical state ", thereby implying the implementation of " social technique ", related thematically to scientific techniques. The " constructors " and the " organizers " among the engineers were to implement this technique of social control and not the " conformist " politicians nor the " bureaucrats "[29]. Much of this rhetoric entered the official ideology of dictatorial regime that dominated Greece from August 1936 to April 1941.

***

During the first forty years of 20th century, Greek engineers succeeded in institutionally consolidating their exclusive ideological representation for technological progress, from which they gained the social and professional profit that extended from the possession of this exclusivity. These social and professional profits consisted of the following.

> - Institutional consolidation was the result of official state policy. The official configuration of a hierarchy of educational and professional institutions guaranteed by the state ensured this success. Supervising these institutions were the Polytechnic School of Athens, which was the sovereign institution of 'applied sciences', and the Technical Chamber of Greece, which was the sovereign professional institution of Greek engineers, and the official technical consultant for the state.

27. " Einai o technikos epistimon dianooumenos ? " (" Is the Technical Scientist Intellectual ? "), *Erga (Works)*, 35 (1926), 237 ; " I Gerousia, oi dianoumenoi kai oi michanikoi " (" The Senate, the Intellectuals, and the Engineers "), *ibid.*, 88 (1929), 467.

28. Thorstein Veblen pushed the logic of the technological systems and the technocratic spirit to extremity. Essentially, he thought that the whole productive system of a country should fall under the absolute control of the engineers. Using the terminology of the Bolsheviks, he spoke for the necessity of engineering soviets. According to this project, engineers should override decisions made by investors and unskilled entrepreneurs, who were regarded parasites on the body of the productive economy motivated by their competitive instincts and their blind addiction to profit. Thorstein Veblen, *The Engineers and the Price System*, reed. Gordon and B, 1971 ; *The Theory of the Leisure Class,* reed. Mendor, 1953 (first ed. 1899) [Greek translation by Giorgos Dalianis, *I Theoria tis Argosholis Taxis,* Athens, Kalvos, 1982].

29. A. Rousopoulos, *Kataskevazein kai Herein (Constructing and Delighting)*, Athens, 1936.

- This institutional consolidation was also gained through the reinforcement of the 'applied sciences', through the identification of engineering knowledge with this type of scientific knowledge, and through the consolidation of their role in designing and implementing technical projects. These enterprises were related instrumentally to the Greek bourgeoisie's anticipation of development of the country and its integration into the modern era.

- On the other hand, the engineers could challenge successfully their wider professional and social demands banking on this gradual amplification of the engineering academic and professional institutions.

The combination of these factors became the vehicle that lead to the emergence of the socio-professional class of Greek engineers, which deserved the same status with the other privileged professional elites such as lawyers, doctors, and military officers. These also provided the framework that led to their domination of the technical professions and set the foundations for their eventual success in achieving the aspiration for a wider social hegemony during the Inter-War period.

ADDENDUM[30]

Table 1

Qualified engineers in 1920

1	Architects	76
2	Civil Engineers	440
3	Mechanical Engineers	104
4	Electrical Engineers	57
5	Chemical Engineers	21
6	Mining Engineers	29
7	Naval Architects	10
8	Surveying Engineers	13
Total		750

30. Sources : *Techniki Epetiris tis Ellados* (*Technical Year Book of Greece*), vol. B' and A' (1934-1935), *op. cit.* ; *Odigos Spoudon E.M.P. 1950* (*Guide for Studies in the Polytechnic School, 1950*), Athens, National Technical University of Athens, 1950, 175. Source : *Ekatontaetiris tou Ethnikou Metsoviou Politechniou, 1837-1937* (*Centennial of the National Technical University of Athens*), Athens, Technical Chamber of Greece [Special edition of *Technika Chronika*, 181 (July 1939)], 132-135.

Table 2

Greek Engineers Graduates from Foreign Technical Schools in 1915[31]

1	Architects	56
2	Civil Engineers	106
3	Mechanical Engineers	29
4	Electrical Engineers	24
5	Chemical Engineers	11
6	Mining Engineers	23
7	Naval Architects	6
8	Surveying Engineers	1
Total		256

Table 3

Employment sections of mechanical engineers, electrical engineers, chemical engineers, mining engineers and naval architects in 1920

Manufacture	28
Electricity	30
Railways	37
Shipyards and Machine shops	23
Mines	19
State and Municipalities	26
Technical Companies	18
Academics	13
Career abroad	27
Total	221

31. These 256 graduates from European technical schools represent only part of the Greek engineers who graduated the last twenty years of 19th century and the first fifteen of the 20th, and they were still alive in 1934. The real number is bigger but there are not still precise rundown elements for them, see *Techniki Epetiris tis Ellados* (*Technical Year Book of Greece*), vol. Β' (1934), *op. cit.*

Table 4

Graduates from the National Technical University of Athens 1889-1940

Year	Civil Engineers	Mechanical and Electrical Engineers	Architects	Chemical Engineers	Surveying Engineers	Total
1889-90	13					13
1890-91	14					14
1891-92	14					14
1892-93	13					13
1893-94	13					13
1894-95	13					13
1895-96	13					13
1896-97	10					10
1897-98	11	3				14
1898-99	4	6				10
1899-00	7	3				10
1900-01	8					8
1901-02	6	1				7
1902-03	9	2				11
1903-04	7	1				8
1904-05	9	7				16
1905-06	13	3				16
1906-07	11					11
1907-08	10	7				17
1908-09	7	2				9
1909-10	16	1				17
1910-11	5	1				6
1911-12	16	2				18
1912-13						
1913-14	13	3				16
1914-15	17	7				24
1915-16	13	5				18
1916-17	16	6				22
1917-18	25	9				34
1918-19	32	8				40

Year	Civil Engineers	Mechanical and Electrical Engineers	Architects	Chemical Engineers	Surveying Engineers	Total
1919-20	26	10		11	1	48
1920-21	30	32	11	2	14	89
1921-22	3	4	2	4	5	18
1922-23	25	14	6	11	12	68
1923-24	6	9	7	13	12	47
1924-25	27	14	10	11	9	71
1925-26	48	18	17	7	14	104
1926-27	39	13	8	10	17	87
1927-28	28	8	6	7	27	76

Year	Civil Engineers	Mechanical and Electrical Engineers	Mechanical Engineers	Architects	Chemical Engineers	Surveying Engineers	Agronomists	Total
1928-29	39	10	Mechanical Engineers	5	7	27	Agronomists	88
1929-30	53	11		4	9	10		87
1930-31	53	10	2	4	10	10		89
1931-32	65	10	2	4	7	8		96
1932-33	52	9	7	17	4	1	4	94
1933-34	64	9	8	16	6	1	5	109
1934-35	71	8	6	29	7		7	128
1935-36	71	13	3	28	10		3	128
1936-37	77	9	6	20	20		2	134
1937-38	55	21	3	8	15		3	105
1938-39	45	18	1	8	9		2	83
1939-40	63	29	9	4	15			120
Total	1298	356	47	214	195	168	26	2304

Table 5

Register of Technical Chamber of Greece in 1939

1	Civil Engineers	1323	48.35%
2	Architects	324	11.79%
3	Electrical Mechanical Engineers	270	9.83%
4	Electrical Engineers	137	4.98%
5	Mechanical Engineers	97	3.54%
6	Chemical Engineers	160	5.84%
7	Mining Engineers	40	1.46%
8	Naval Architects	18	0.65%
9	Surveying Engineers	257	9.39%
10	Engineers having more specialties than one	41	1.49%
11	Civil Engineers who worked as Architects	40	1.46%
12	Civil Engineers who worked as Surveying Engineers	29	1.05%
Total		2736	

Table 6

Greek engineers, who graduated from foreign technical schools and they were members of the Technical Chamber of Greece, in 1939

1	Belgium	128
2	G. Britain	29
3	France	187
4	Germany	225
5	Switzerland	130
6	U.S.A.	23
7	Italy	43
8	Turkey[32]	5
9	Russia	19
Total		789

32. The Greek Engineers from Turkey were mainly alumni of School of Beaux Arts of Istanbul and all of them graduated before 1922.

Table 7

National Technical University of Athens Average of Graduation 1889-1939

Decade	Year Average
1889-1899	12,7
1899-1909	11,3
1909-1919	19,5
1919-1929	69,6
1929-1939	105,3

Table 8

Members of Technical Chamber of Greece in 1935

Speciality	Category[33]	Athens - Piraeus			-	County			-	Abroad	Part. total	Gen. total
Civil Engineers	-	13									13	
	-	138				143					281	
	C		162		683		118		377		280	1070
	D			370				116		10	496	
Architects	-										-	
	-	37				10					47	
	C		21		170		11		51		32	223
	D			112				30		2	144	
Surveying Engineers	-										-	
	-	74				23					97	
	-		18		142		16		54		34	198
	-			50				15		2	67	
Mechanical and Electrical Engineers	-	10									10	
	-	60				10					70	
	C		143		338		66		96		209	440
	D			125				20		6	151	
Chemical Engineers	-	1									1	
	-	24				7					31	
	C		28		102		20		35		48	140
	D			49				8		3	60	

33. Category A : Professors of National Technical University of Athens

Category B : State employees

Category C : Employees in private sector, municipalities etc

Category D : Professionals

Source : *Techniki Epetiris tis Ellados* (*Technical Year Book of Greece*), vol. B' and A', Athens, Technical Chamber of Greece, 1934-1935.

Speciality	Category	Athens - Piraeus					County				Abroad	Part. total	Gen. total
Mining Engineers	-	1										1	
	-		9				1					10	
	C			8		25		8		12	1	17	39
	D				7				3		1	11	
Naval Architects	-	-					-					-	
	-		9				-					9	
	C			2		12		-		-	-	2	12
	D				1				-		-	1	
Engineers having more than one specialties	-	2										2	
	-		4				-					4	
	C			4		20		1		3	-	5	24
	D				10				3		-	13	
Number of members by category	-	27										27	
	-		355				194					549	
	C			386		1492		240		629	1	627	2146
	D				724				195		24	943	

INSTITUTIONALISATION OF SCIENCE EDUCATION AND SCIENTIFIC RESEARCH IN TURKEY IN THE 20ᵗʰ CENTURY

Sevtap KADIOGLU

The " scientific life " of a society is characterized by the activities of certain scientific institutions in that society, the universities playing the leading role in this process. Thus, one way of evaluating the scientific life of a community is to study closely the activities and functions of those institutions where scientific research is conducted. Considering this fact, the present paper aims to analyse and shed light on the activities of two pioneering institutions of science in Turkey that illuminate the history of the institutionalisation of education and research in exact sciences in the 20ᵗʰ century. The first institution under study is the Istanbul University Faculty of Science, the foundation of which dated back to 1900s and which was the only institute for decades to carry out education and research in exact sciences. The other one is the Scientific and Technical Research Council of Turkey (TÜBITAK), an organisation that differs from educational institutions or private research foundations in the sense that it is a state organisation conducting research in exact and applied sciences as well as supporting, encouraging and guiding the research activities of other individuals and organisations. A detailed study of the activities of these two institutions will clarify how scientific education and research have been organized in Turkey.

During the classical period of Ottoman Turkey, that is up to the beginning of the eighteenth century, the *madrasas* were the primary institutions where mathematics and natural sciences were taught. Starting from the second half of the eighteenth century, new educational institutions were founded upon Western models to transfer modern European sciences into Turkey. Mathematical and natural sciences then became included into the curricula of these newly founded institutions, the first of which were the Imperial School of Naval Engineering (1773) and the Imperial School of Military Engineering (1795).

Apart from these institutions came the idea to establish a *Darülfünun* (University) in the mid-nineteenth century. The initial attempts to establish the Darülfünun can be traced back to 1846 whereas the actual foundation came about in 1900, just at the turn of the century. Mathematics and sciences were taught in the form of conference lectures in the first attempt of 1863. Though these conference lectures arose great interest among the students, this first attempt could not attain the desired success and did not last long since it was not originally designed as a form of higher education. The second Darülfünun of 1870, called the Ottoman Darülfünun (*Darülfünun-ı Osmani*) was set off with a much different and more mature understanding ; however, this second attempt, too, was a failure as result of various difficulties and obstacles, and the instruction of mathematics and science was once more interrupted. These courses were taught after a while within the School of Civil Engineering, founded as part of the third Otttoman Darülfünun of 1874, to constitute basis for engineering education.

FACULTY OF SCIENCE, ISTANBUL UNIVERSITY

As already stated, the teaching of mathematics and sciences had never been continuous in the first three Darülfünun attempts. A continuous instruction was finally carried out in the Department of Mathematical and Natural Sciences within the Imperial Darülfünun founded in 1900, though the political conditions of the period were rather tough and opportunities were limited. This new institution laid the foundations for the present day Istanbul University Faculty of Science and the teaching of mathematics and natural sciences was finally institutionalized here as independent disciplines.

The first graduates received their degrees in 1904 and certain measures were introduced to elevate the level of education before the First World War (The Regulations of 1912). Not many students were attracted to this faculty, however, since the graduates had more restricted job opportunities than those of other faculties.

The years of the First World War witnessed the arrival of German scholars at the Faculty of Science, starting a period of restoration and development. As a productive result of the alliance between Germany and the Ottoman State during World War I, research and education facilities were strengthened in the Darülfünun by these German professors. Twenty scholars signed 5-year-term contracts starting from the academic year 1915-1916 on and six of them were assigned to teach at the Faculty of Science : three in chemistry, others in geology, zoology and botany, with Turkish assistants to help them. These scholars were :

Prof. Dr. Boris Zarnick (Zoology)

Ass. Prof. Dr. Walther Penck (Geology and Geography)

Ass. Pof. Dr. Leick (Botany)

Ass. Prof. Dr. Fritz Arndt (Inorganic Chemistry)

Assistant Dr. Kurt Hoesch (Organic Chemistry)

Assistant Dr. Gustav Fester (Industrial Chemistry)

When the war ended, German scholars expressed their wish to return to their home country although their terms of contract had not ended yet. That Ottoman rulers offered a rise in their salaries did not make them change their mind and the Germans left their offices just after the war ; the courses were then taught by their Turkish assistants. This act of bringing foreign scholars from abroad, usually referred as the " first partial reform at the Darülfünun " in Turkish higher education, thus proved ineffective, not attaining the anticipated success. Several factors were influential in the failure of the plan : the scholars did not know any Turkish, and thus could not lecture in Turkish though they had previously agreed to do so ; the number of students was rather low due to the ongoing war etc. The endeavours of these scholars, however, in the establishment of the Yerebatan Institute of Chemistry and the Institute of Geology in Vefa as well as their later efforts to obtain the necessary instruments for these institutes are rather noteworthy. When the German scholars returned to their countries before rendering the faculty fully active, new regulations were devised to bring a new administrative and scientific organisation to the Darülfünun. A new course schedule was introduced where some courses were abolished and others were introduced. Those were hard times for the Darülfünun, just as everything else in the country.

The first few years following the foundation of the Turkish Republic witnessed the launching of a P.C.N. (Physics, Chemistry, Natural Sciences) class and the bestowal of " Teaching Certificates " to the graduates of the Faculty of Science that would enable them to teach at secondary schools. Certificates for Chemical Engineering and Mechanical-Electrical Engineering were also handed out during this same period, allowing the Faculty of Science to contribute to engineering education.

In 1926 French scholars were brought to the Darülfünun Faculty of Science with the intent to promote the level of education. This was called the second partial reform in higher education. Why was France chosen this time ? Since a cultural agreement had been signed with this country during this time. Initially 5 scholars in mathematics, physics, physical-chemistry, zoology and electro-mechanics had arrived from France ; yet the number gradually increased the following years. These first five scholars were Mentere for mathematics, Fleury for physics, Hovasse for zoology, Faillebin for the newly founded discipline of physical-chemistry, and Duscio for the newly founded electro-mechanics.

Neither the regulations of 1912 and 1919 nor the partial reforms of 1914 and 1926 could prevent the government from issuing the University Reform of 1933 which brought radical changes in all aspects of higher education.

The reform of 1933 was a landmark in the history of Turkish higher education where the Darülfünun was abolished and Istanbul University was officially founded. Just before the reform, Prof. Albert Malche from the University of Geneva was invited in 1931 by the Turkish government to prepare a report on the reform that was about to take place in the Darülfünun. Prof. Malche submitted his report titled " Report on the Istanbul Darülfünun " to the Ministry of Education 1 June 1932, where he brought various criticisms and recommendations to the Darülfünun.

Malche's report emphasized some significant aspects and defects of the activities of the Faculty of Science : there was no independent education in biology. The Chair of Natural Geography and the Chair of Anthropology within the Faculty of Medicine should both be moved into the Faculty of Science since these disciplines had a direct relation to natural sciences. Studies and research at the zoology station had to be developed and matured. The diplomas granted by the Institute of Electro-Mechanics should be equivalent to those of the Engineering School. Laboratories of related disciplines should mutually make use of their instruments so that no new tools were needed for these labs. The library was messy and needed to be arranged. Lastly, a Ph.D. class was needed within the Faculty of Science.

Once the preparations of a reform were completed, the Darülfünun was closed down on 31 July 1933 and Istanbul University was founded as from 1 August 1933. The reform brought along a large scaled dismissal of scholars and instructors ; only one third of the Darülfünun instructors were employed in the newly founded university. All the teaching staff at the Faculty of Science were dismissed except three. These three were Ali Yar from the chair of mathematics, Hamit Nafiz [Pamir] from the chair of geology and Fahir [Yeniçay] from the chair of physics.

The fact that there was no sound and definite criteria for the mass dismissal of these scholars has been criticized ; it is also noteworthy that the Faculty of Science, which was actually the least criticized faculty in Malche's report, was the place where this elimination process took place at most.

After the dismissal, the teaching staff were reinforced by German scholars of Jewish origin fleeing from the Nazi regime in their country as well as young Turkish scholars who had completed their undergraduate studies abroad and returned home at that time. All chair directors were foreign scholars except for geology, a fact indicating that all educational activities in the Faculty of Science were conducted and administered by the Germans until 1950s. Chair directorships were later passed on to the Turkish assistants. Between 1933-1946, generally referred as the " golden age " of the Faculty of Science, a total of 41 doctorates were completed : 10 in mathematics, 2 in astronomy, 2 in physics, 8 in chemistry, 8 in botany, 6 in zoology, 5 in geology, whereas at the time of the Darülfünun, Ph.D. degrees could only be attained through the personal initiatives and efforts of individual scholars. One can thus comprehend that research activities were intensified at the Faculty of Science only after the 1933 University Reformation.

The Zeynep Hanım Mansion, where the Faculty of Science had been housed since the Second Constitutional Period, was burnt down in 1942 and this event undoubtedly had a negative influence on the research and educational activities of the faculty. The fire not only destroyed all tools and instruments within the building, but also forced the faculty to locate at various premises within the city. The later University Law of 1946 altered the whole educational system of the faculty ; the certificate system was abandoned for the branch and course system.

To sum up, mathematics and natural sciences were taught in all three attempts to establish a Darülfünun, but the core of today's Istanbul University Faculty of Science was the Department of Mathematical and Natural Sciences founded in 1900 within the Darülfünun. This institution underwent various structural rearrangements and had a final organisation with the University Reform of 1933. The Faculty of Science may be said to have constituted the basis for other faculties of science in Turkey and has been the first institution to start an independent education and research tradition in exact sciences.

THE SCIENTIFIC AND TECHNICAL RESEARCH COUNCIL OF TURKEY (TÜBITAK)

In the first half of the twentieth century, that is the end of the Ottoman period and beginning of the Republican period, there were also other institutions in Turkey doing research in health sciences, agriculture, stockbreeding, mining etc. and founded to meet the practical needs of the state and public. Among these organizations were the General Directorate of Mineral Research and Exploration, General Directorate of Electrical Power Resources, Survey and Development Administration, Etibank, General Directorate of State Hydraulic Works, and the Atomic Energy Agency of Turkey. Yet, each of these research institutions was directed towards some specific aim, and neither carried the mission of conducting general research in the field of exact sciences. In this respect, TÜBITAK has been the first and only institution founded for this end. The contribution and significance of exact sciences to the lives of nations became so apparent after the Second World War that European and American scientists considered it the responsibility of all governments to improve this field of research. Researching in exact sciences was a far too important task to be left to the individual efforts of institutes or universities alone ; therefore all work undertaken in this area should be coordinated by the state. As a result of this policy, national research organisations were founded in almost all Western countries, and the agenda was also set for a similar institution in Turkey. The initial attempts came from UNESCO National Commission in Turkey in 1953, which proved unfruitful. New attempts were started in 1960 and reports by two important international organisations the same year contributed to the foundation of TÜBITAK. The reports were submitted in the NATO Commission of Science and UNESCO Regional Conference, suggesting the establishment of a national research institute to conduct scientific and technical research in Turkey.

Firstly, a regional conference was organized by the UNESCO Cooperation Center for Middle Eastern Science on 19-22 December 1960 in Cairo to encourage scientific research in Middle Eastern countries and provide cooperation on this issue. Turkey was also invited ; Turkey's participation in such a conference was an important step in paving the way for the foundation TÜBITAK. United Arab Republic (Egypt and Syria), Iraq, Jordan, Lebanon, Morocco, Tunisia and Sudan were the other countries that participated in the conference. The most important decision taken was that scientific research in the modern world was not peculiar to individuals but should be a collective activity of the economic, industrial and social life of a society. Thus, it was necessary for the countries to establish national councils, apart from universities, supporting, encouraging and coordinating scientific research.

Secondly, the NATO Committee of Science Research Group, which came together to consider the means to improve exact sciences in NATO member countries and which consisted of many well-known scientists, passed a report in 1960 stating that all western countries had to have one or more science councils or organizations supporting the exact and applied sciences and working to establish a certain quality in the exact sciences. These institutions were to have executive councils with authority and the scientists within these organizations had to employ scientists with abilities to perform the necessary deeds.

These two reports prepared by two important international organizations — the report of the NATO Commission of Science and the report by UNESCO Regional Conference — had a great impact on the attempts to establish TÜBITAK.

On the other hand, Turkish scholars and scientists presented many views and had many publications on expanding sciences to all classes of the society, raising scientists in Turkey, and the foundation of a research institution. In their publications the scientists emphasized the fact that in order to attain an economic development, a country should have an advanced level in exact sciences. They also supported the idea of establishing a central organization encouraging and coordinating the research and development activities in the country.

The First Developmental Plan of Turkey also laid the foundations for such a national research organization. The main objective of the First Five-Year Development Plan (1963-1967) was to benefit from exact and applied sciences for the solutions of the problems and obstacles in Turkey's development and progress. The main fact preventing the development of research activities was the lack of an institution organizing and coordinating the rather dispersed and unorganized research organizations and their activities. Necessary measures had to be taken to direct and guide the political activities in Turkey about all kinds of research in social and exact sciences and in technology.

Thus, the idea of founding a national research institution was refined and also found favour amongst scientists ; as a result, high officers of the army, who were then running the government, initiated the attempts and TÜBITAK was finally founded at the end of 1963, being the first national research institute put in charge of promoting, developing, organizing and coordinating research and development in the fields of exact sciences.

Before the actual foundation of TÜBITAK, similar institutions worldwide were examined and adopted as models. The organisational structure at the foundation consisted of such organs as the Science Board, Consultancy Committee, General Secreteriat and Research Groups. This newly founded institution was under the fold of the highest executive organ of the government, the Prime Ministry, and yet had adequate administrative (*i.e.* election system) and financial autonomy. All the members of the Science Board, the main administrative organ and decision making body, were elected by other members of this committee.

The first research groups established within TÜBITAK by the foundation law were the Mathematics, Physics and Biological Sciences Research Group, Engineering Research Group, Medical Research Group, Veterinary and Stock-Breeding Research Group, Agriculture and Forestry Research Group, and the Group for Training [Young] Scientists. The Science Board was also endowed with the authority to establish new research groups as needs arose.

The institution carried on its activities for 25 years in this way and finally a decree in 1987 brought radical administrative changes in this respect which abolished the authority of the Science Board to elect its own members ; and the administrators were now appointed by the Prime Minister himself. The decree also laid down that all research and development activities should be geared to the targets of the five-year economic development plans of the country and be open to public competition. The decree also emphasized the significance of TÜBITAK as an organisation meeting the research demands of official and private organisations.

Though this new formation decreased the administrative autonomy of TÜBITAK in many respects, it was an important step in the sense that it promoted the TÜBITAK-industry relations. Six years later a new system of autonomy was introduced with the newly reformed TÜBITAK Law of 1993. The original system where the members of the Science Board were elected by the council itself was restored by this law and the administrative autonomy of the institution was reestablished and enhanced, but the Chairman of the Science Board and the President of TÜBITAK remained the same person, as in the decree of 1987. The President, as the chairman of the Science Board implemented its decisions and was the head of the entire administration. This new organisational status might be considered a synthesis of two distinct laws : it restored the autonomy granted by the Law of Foundation and preserved the presidency system of the 1987 decree.

As for the activities of TÜBITAK, its main task is to conduct research in-house and in its institutes and back up other academic and industrial projects through its Research Groups. By this way TÜBITAK fulfills its founding function of " supporting, encouraging and coordinating scientific research ". The majority of TÜBITAK's research are carried out at an institution known as Marmara Research Center (MAM), founded in 1972. There are also other similar organs, such as Defense Industries Research and Development (R&D) Institute founded in 1972 for military purposes ; and Information Technologies and Electronics Research Institute in Ankara founded in 1982 for research in electrical-electronical sciences. TÜBITAK's various research groups in exact sciences, engineering, medicine, veterinary and stockbreeding, agriculture and forestry, environmental issues etc. carry out its task of supporting and encouraging research in other institutions.

With a more detailed and closer study of TÜBITAK's activities, one can figure out that it not only supports science and technology projects but also fulfills certain very important tasks for the development of science, such as publishing scientific journals, as well as books and monthly popular science magazines that make science accessible to the public ; supporting scientists and researchers with awards ; implementing tasks undertaken through international scientific and technical cooperation agreements. It also supports R&D

activities and innovations in industry and has been involved in the task of determining Turkey's science and technology policies since 1993.

Since its foundation, TÜBITAK has sought to train young researchers, popularize subjects of science and research, and support all research activities ; by this way it has contributed greatly to the scientific life of Turkey. After 1980s TÜBITAK became an institution largely committed to promoting cooperation with industry. In 1990s, on the other hand, it took on another task where its role in determining and shaping the country's science and technology policy became much more preeminent.

It is a generally acknowledged fact that a country's ranking in international lists of publications is an important criteria in the evaluation and assessment of its scientific life and activities. While at the end of 1980s Turkey was ranked 41st amongst the countries which had most publications in sciences as found in the Index of Scientific Citations, it went up to the 29th place in 1996 and to the 27th in 1997 and was ranked 25th between 1998 and 2001, a success in which TÜBITAK has undoubtedly had an important share as a policy making institution on science and technology.

When TÜBITAK was first founded, Turkey had only a few universities and some 20 research institutions and laboratories, mostly working on agriculture and had really limited facilities for science and technology related research. Founded in such a system, the main target of TÜBITAK was to train researchers and support the research activities in all areas. Priorities were also established ; for example, improving the production in industry and directing the research activities to this end was one of the most significant of them. However, the industry was still in its infancy and there was no demand for research and development activities. Thus, almost all the research funds of TÜBITAK went into basic academic research and investigations, i.e. to the universities. The main contribution of TÜBITAK remained as training researchers and encouraging talented and successful students to get into research activities.

In 1980s, parallel to the developments in the world, Turkey gave up importing industrial policies ; with the changes of 1987 in TÜBITAK Law, the organization started to serve intensely the needs of the industry through its research activities. As a result, the relations of TÜBITAK with the industry were strengthened.

With the foundation of TÜBITAK, the government for the first time took up all works in science and technology as a whole. Its foundation might be considered as the practical application of a certain science policy in Turkey for the first time. TÜBITAK maintains its distinctive place in Turkish science as an institution conducting, supporting and encouraging research activities, and especially after 1993, as an institution actively taking part in the making of science and technology policies.

To conclude our paper on these two leading institutions, one can say that the first institution to start the tradition of independent research and education in exact sciences was the Istanbul University Faculty of Science. The scholars and scientists educated here constituted the staff of later faculties of science in Turkey and carried out the task of doing research. Some of those scholars later took part in the foundation of TÜBITAK, established to train young researchers, disseminate the concepts of science and research, and finally support and encourage scientific research. Thus, both institutions and the scholars trained here contributed greatly to the institutionalisation of science education and scientific research in Turkey.

BIBLIOGRAPHY

- *Başbakanlık Devlet Planlama Kurulu, Kalkınma Planı (Birinci Beş Yıl) 1963-1967*, Ankara, Başbakanlık Devlet Matbaası, 1963, 466-467.

- *Bilim ve Teknoloji Mevzuatı*, Ankara, 1993.

- *Bilim ve Teknoloji Yönetim Sistemleri : Ülke Modelleri ve Türkiye*, Ankara, TÜBITAK, 1996.

- " 60. Yılında TÜBITAK 20 Yaşında ", *Bilim ve Teknik*, VIL, vol. 16, issue 192 (November 1983), 1-2.

- Hirsch, Ernest, *Dünya Üniversiteleri ve Türkiye'de Üniversitelerin Gelişmesi*, Ankara, 1950.

- Ihsanoğlu, Ekmeleddin, " The Genesis of " Darulfünun " : An Overview of Attempts to Establish the First Ottoman University ", in Daniel Panzac (ed.), *Histoire Économique et Sociale de l'Empire Ottoman et de la Turquie (1326-1960)* : Actes du Sixième Congrès International Tenu à Aix-en-Provence du 1er au 4 Juillet 1992, Paris, Peeters, 1995, 827-842.

- Ihsanoğlu, Ekmeleddin (ed.), *History of the Ottoman State, Society and Civilisation*, vol. 2, Istanbul, IRCICA, 2002.

- Ihsanoğlu, Ekmeleddin, " Darülfünun Tarihçesine Giriş, Ilk Iki Teşebbüs ", *Belleten*, LIV/210 (1990), 669-738.

- Ihsanoğlu, Ekmeleddin, " Darülfünun Tarihçesine Giriş (II), Üçüncü Teşebbüs : Darülfünun-ı Sultani ", *Belleten*, LVII/218, 201-239.

- Ishakoğlu, Sevtap, " 1900-1946 Yılları Arasında, Darülfünun ve Istanbul Üniversitesi Fen Fakültesi'nde Matematik ve Fen Bilimleri Eğitimi ", in Feza Günergun (ed.), *Osmanlı Bilimi Araştırmaları*, Istanbul, 1995, 227-283.

- Ishakoğlu Kadıoğlu, Sevtap, *Istanbul Üniversitesi Fen Fakültesi Tarihçesi (1900-1946)*, Istanbul, 1998.

- Ishakoğlu Kadıoğlu, Sevtap, " Türkiye Bilimsel ve Teknik Araştırma Kurumu'nun (TÜBITAK) Kuruluşu ve Türk Bilim Hayatındaki Yeri ", Istanbul 1999, unpublished Ph.D. thesis.

- Ishakoğlu Kadıoğlu, Sevtap, " 1900-1946 Yılları Arasında Darülfünun ve Istanbul Üniversitesi Fen Fakültesi'nde Kimya Eğitimi ", in Emre Dölen (ed.), *Istanbul Üniversitesi Kimya Fakültesi' nin Kuruluşunun Otuzuncu Yılı*, Istanbul 1997, 19-58.

- Ishakoğlu, Sevtap, " 1900-1946 Yılları Arasında Darülfünun ve Istanbul Üniversitesi Fen Fakültesi'nde Botanik, Zooloji ve Jeoloji Eğitimi ", in Feza Günergun (ed.), *Osmanlı Bilimi Araştırmaları II*, Istanbul, 1998, 319-348.

- Ishakoğlu Kadıoğlu, Sevtap, " The Teaching of Mathematical and Natural Sciences at the Darülfünun and Istanbul University Faculty of Science ", in E. Ihsanoğlu & F. Günergun (eds), *Science in Islamic Civilisation* : Proceedings of the international symposia " Science institutions in Islamic Civilisation & Science and Technology in the Turkish and Islamic World ", Istanbul, 2000, 65-75.

- " Kahire Konferansı'na Ait Rapor", UNESCO Türkiye Milli Komisyonu, 1961.

- Kocatürk, Utkan, " Atatürk'ün Üniversite Reformu Ile Ilgili Notları ", *Atatürk Araştırma Merkezi Dergisi*, issue 1 (1984), 1-94.

- *OECD Türkiye Ulusal Bilim ve Teknoloji Politikası Raporu*, Ankara, 1996.

- Pak, Namık Kemal, " Türkiye Cumhuriyeti'nin 75.Yılında Türkiye Bilimsel ve Teknik Araştırma Kurumu'nun (TÜBITAK) Doğuşu, Evrimi ve Politikaları ", *Türkiye Cumhuriyeti' nin Kuruluşu' nun 75. Yılında Bilim (Bilanço : 1923-1998) Ulusal Toplantısı*, Book 2, Vol. I, Ankara, TUBA Publications, 1999.

- Tunçay, Mete - Haldun Özen, " 1933 Dârülfünûn Tasfiyesi veya Bir Tek Parti Politikacısının Önlenemez Yükselişi ve Düşüşü ", *Tarih ve Toplum* (October 1984), 6-20.

- TÜBITAK'ın 1980-1996 Tarihleri Arasındaki Yıllık Faaliyet Raporları.

- " Türkiye Bilimsel ve Teknik Araştırma Kurumu Kurulması Hakkında Kanun Tasarısı Gerekçesi ", *Millet Meclisi Tutanak Dergisi*, vol. 16 (8.4.1963), 68. Birleşim, Dönem 1, Toplantı 3, 4.

- Widmann, Horst, *Atatürk Üniversite Reformu*, translated by Aykut Kazancıgil - Serpil Bozkurt, Istanbul, 1981.

- Yeniçay, Fahir, " Istanbul Üniversitesinde Fiziğin Gelişmesi ", *Istanbul Üniversitesi Fen Fakültesi' nde Çeşitli Fen Bilimi Dallarının Cumhuriyet Dönemindeki Gelişmesi ve Milletlerarası Bilime Katkısı*, Istanbul, 1982, 36-53.

CONTRIBUTORS

Gábor ÁGOSTON
Georgetown University
Washington DC (USA)

Yiannis ANTONIOU
National Technical University of
Athens
Athènes (Grèce)

Fotini ASSIMACOPOULOU
Research Institute of Modern Greece
of the Academy of Athens
Athènes (Grèce)

Michalis ASSIMAKOPOULOS
National Technical University of
Athens
Athènes (Grèce)

Cemil AYDIN
Harvard Academy for International
and Area Studies
Cambridge, MA (USA)

Yakup BEKTAŞ
Duke University
Durham, North Carolina (USA)

Atilla BIR
Istanbul Technical University
Istanbul (Turkey)

Sonja BRENTJES
University of Oklahoma
Oklahoma, OK (USA)

Rainer BRÖMER
University of Aberdeen
Aberdeen (United Kingdom)

Konstantinos CHATZIS
École Nationale des Ponts et
Chaussées
Paris (France)

Ekmeleddin IHSANOGLU
Research Center for Islamic History,
Art and Culture (IRCICA)
Istanbul (Turkey)

Mustafa KAÇAR
Istanbul University
Istanbul (Turkey)

Sevtap KADIOGLU
Istanbul University
Istanbul (Turkey)

Peter MENTZEL
Utah State University
Logan, UT (USA)

Nathalie MONTEL
École Nationale des Ponts et
Chaussées
Paris (France)

Efthymios NICOLAÏDIS
National Hellenic Research
Foundation
Athènes (Grèce)

Antoine PICON
Harvard University
Cambridge, MA (USA)

Yakov M. RABKIN
Université de Montréal
Montréal (Canada)

Maria TERDIMOU
National Hellenic Research
Foundation
Athènes (Grèce)

Ioli VINGOPOULOU
National Hellenic Research
Foundation
Athènes (Grèce)